ASP.NET Core 真机拆解

罗志超　著

ndName": "IISExpress",
Browser": true,
mentVariables": {
NETCORE_ENVIRO
ndName": "Project",
Browser": true,
tionUrl": "http://localhost:5000",
mentVariables": {
uthenticatior
sAuthenti
onUrl": "h ost:68425",
"sslPort": 0
"IIS Ex
dName": "IISExpress",
rowser": true,
nentVariables": {
ETCORE_ENVIRONMENT": "D

人 民 邮 电 出 版 社
北 京

图书在版编目（CIP）数据

ASP.NET Core真机拆解 / 罗志超著. -- 北京 : 人民邮电出版社, 2020.10
ISBN 978-7-115-54048-5

Ⅰ. ①A… Ⅱ. ①罗… Ⅲ. ①网页制作工具－程序设计 Ⅳ. ①TP393.092.2

中国版本图书馆CIP数据核字(2020)第084130号

内 容 提 要

对于读者来说，学习 ASP.NET Core 就像学习使用一部新手机。手机内部组成结构可能有改变，各个元器件的性能可能有提升，元器件之间的兼容性可能更好。但在操作方式上，用户除了要学习如何使用个别的新功能以及适应系统更快的反应速度外，并没有太多改变。

基于以上，本书内容主要分为三部分。第一部分是 ASP.NET Core 的使用说明，简要介绍如何使用 ASP.NET Core 创建项目及其与 ASP.NET 4.x 的不同点；第二部分是真机拆解，讲解 ASP.NET Core 框架的内部运行逻辑；第三部分是通过一个项目案例回顾全书知识点，并介绍如何在 ASP.NET Core 中对用户进行认证和授权。

本书适用于 ASP.NET 开发从业者以及想要学习 ASP.NET Core 的人员，要求读者具有 ASP.NET MVC 基础。

◆ 著　　　罗志超
责任编辑　张　爽
责任印制　王　郁　焦志炜

◆ 人民邮电出版社出版发行　　北京市丰台区成寿寺路 11 号
邮编　100164　　电子邮件　315@ptpress.com.cn
网址　https://www.ptpress.com.cn
固安县铭成印刷有限公司印刷

◆ 开本：800×1000　1/16
印张：19　　　　　　2020 年 10 月第 1 版
字数：420 千字　　　2024 年 7 月河北第 3 次印刷

定价：69.00 元

读者服务热线：(010)81055410　印装质量热线：(010)81055316
反盗版热线：(010)81055315
广告经营许可证：京东市监广登字 20170147 号

前　言

编写背景

2016 年，.NET Core 1.0 发布时，它主打的跨平台和高性能特性吸引了许多人，相信许多从事.NET 工作的人都很激动。但因为它刚发布，许多人都只是在观望，毕竟一项新技术从出现到应用于生产环境要经历一个较长的过程，需要“踩坑”，需要第三方框架的支持。此外还有时间成本的问题，现在各种新技术涌现得越来越快，只有不断地学习才能不掉队，而这给人的感觉就是时间越来越不够用。

直到 2018 年初，.NET Core 2.0 已经发布有一段时间了（2.0 版在 2017 年 8 月发布），有人试着将其使用在生产环境中，反馈还不错，于是我也决定在这一年逐步在新项目中使用 ASP.NET Core。ASP.NET Core 主要面向 Web 和 API 开发，在跨平台方面难度较小，发展很快，已经非常成熟；而 ASP.NET Core 对 WinForm 和 WPF 的支持似乎是相对滞后的，这主要在设计器部分表现得比较明显。在学习和使用的过程中，我将学习笔记发布到了博客园，受到了许多园友的关注，并且有多篇文章上了“编辑推荐”，这是我之前没有想到的，给了我很大的鼓励。

在 ASP.NET Core 的使用过程中，大多数人很自然地会将它和原有的 ASP.NET MVC 进行比较，二者在使用方法上有很多相似之处。因为 ASP.NET Core 在 GitHub 上是开源的，所以我也逐步养成了读源码的习惯。一边读源码，一边用一些小例子做验证，并按照自己的理解整理成笔记分享到博客园。博客园中有很多人无私分享，每一个.NET 人都希望.NET 技术和生态能有更好的发展，我也希望我能尽一份绵薄之力。

后来出版社的张爽编辑问我是否考虑把这些内容汇编成书，我感到非常惊讶，因为从没想过写书这件事能和自己有关系。对于我来说，写实体书和在网上发博客文章的区别太大了，我的博文偏口语化，而且对于一些简单的、似乎是“大家都懂的”知识点就一带而过了，经常有“跳跃”；而在写作实体书时要非常注意语法和措辞，在此感谢出版社的编辑们的耐心指导和纠正。

现在，.NET Core 3.0 已经发布了，本书内容也是基于 3.0 版本的。目前许多公司开始打算或者已经将 ASP.NET Core 应用于生产环境了，也有很多人准备转向 ASP.NET Core。我把博客中的内容整理出来，希望能对像我一样希望深入学习 ASP.NET Core 的朋友们有所帮助。

学习建议

本书主要借助一些例子配合 ASP.NET Core 的源码进行讲解，建议读者打开源码进行调试，跟着例子试一试，这样做有两个好处。

（1）**熟悉架构内部的处理流程**。从一个小例子开始，就像调试自己的代码一样，看看 ASP.NET Core 的内部架构是如何运作的，便于读者理解知识点并加深印象。

（2）**学习 ASP.NET Core 的架构设计模式**。在源码学习的过程中多思考，想想为什么要这样设计？好处是什么？用到了哪些设计模式？哪些地方是用来方便我们使用的“脚手架”？哪些地方是预留给我们做扩展的？整套源代码是很庞大的，不求完全“吃透”，但大体学习下来还是会让人受益匪浅的。

在平时的学习过程中，有以下两点建议。

（1）**多实践，多积累**。可以设计一个实用的小项目，尽可能地涉及自己正在学习和近期想要学习的技术。不要担心没学过或没用过，可以参考网上分享的例子或者读一些技术书籍。不求尽快完成，但求每天完成一点，一个个问题被解决的同时，也会积累一些经验。

（2）**多交流，多分享**。例如写博文、参与技术交流、参加技术沙龙等，与其他同行一起学习和讨论。在这个过程中，你可能会发现对于某些问题，别人原来有更好的处理方法，并且别人也可以帮助我们指出自己的不足。

本书特色

（1）**系统框架分析**。本书不是一本手把手介绍基础操作的书，基础应用部分较少，不会讲什么是 Controller、什么是 Action 等。侧重于 ASP.NET Core 框架的内部运行机制分析，使读者在懂得使用的基础上，深入了解系统的框架结构、运行机制，做到“知其然，并知其所以然”。希望本书的读者有 MVC 基础，若没有，可参考一些基础的 MVC 方面的书籍。

（2）**通俗易懂不枯燥**。本书是作者对 ASP.NET Core 进行深度学习后，通过自己对框架的理解来进行描述的，并非对源码的枯燥注释，也不是对外文书籍的生硬翻译。

（3）**结合案例**。本书大部分内容是结合实际案例进行介绍的，且案例不限于代码，还会将部分逻辑类比工作、生活、小说故事等进行讲解，使内容通俗易懂。此外，本书的最后一章会通过一个综合案例回顾书中提到的知识点。

本书结构

下面介绍本书的结构设计思路，希望读者能先看一下，这样会对理解本书内容有很大的帮助。

第一部分：从使用角度介绍 ASP.NET Core。

第 1 章介绍为什么要使用 ASP.NET Core，它的优势是什么，以及它为什么能跨平台。

第 2 章介绍开发环境的准备工作，Visual Studio 在不同环境下的安装过程。

第 3 章新建一个项目，逐一介绍项目中各个文件夹和文件的用途，从整体的角度了解项目的构成。

第 4 章介绍_Layout 和_Viewstart，了解页面的加载顺序。

第 5 章介绍 ASP.NET Core 的新成员——TagHelper。

第 6 章通过一个例子介绍如何将项目部署到 CentOS，包括安装 Nginx、SSL 免费证书申请、多网站及证书配置等。

第二部分：解析 ASP.NET Core 框架的内部处理机制。

第 7～14 章：介绍应用启动过程中的准备内容。

第 7 章从宏观的角度介绍 ASP.NET Core 的运行机制以及 ASP.NET Core Application 的架构等。

第 8 章介绍应用启动过程中做的准备工作，一些关键组件是如何被构建、启动的。

第 9 章介绍后台服务的应用场景是什么，以及实现方式和注意事项等。

第 10 章介绍 ASP.NET Core 的依赖注入。在 ASP.NET Core 中，绝大部分组件是通过依赖注入提供的。

第 11 章介绍 ASP.NET Core 的内置日志、注意事项以及配置简要说明等。

第 12 章介绍应用的常见配置方式以及系统框架的内部处理机制。

第 13 章介绍另一种配置方式，即 Options 模式的使用及内部处理机制。

第 14 章介绍 ASP.NET Core 的请求处理管道，这是 ASP.NET Core 的核心内容之一。

第 15～20 章：介绍应用启动后，收到用户的请求后是如何处理并返回结果的。

第 15 章介绍 ASP.NET Core 对静态文件的访问和授权等。

第 16 章介绍路由的配置等。

第 17 章介绍 Action 是如何被执行的。

第 18 章介绍请求的参数是如何与 Action 中的参数一一绑定的，不同类型的参数是如何被处理的。

第 19 章介绍各种 Filter 的用法，以及 Filter 的获取和执行等。

第 20 章介绍如何控制返回数据的类型及其内部处理机制等。

第三部分：项目案例

第 21 章通过一个项目案例回顾本书涉及的知识，该案例是一个由 ASP.NET Core、微信小程序和 MongoDB 组成的项目，并介绍如何在 ASP.NET Core 中对用户进行认证和授权、Swagger 的使用等。

建议和反馈

写一本书是一项极其琐碎、繁重的工作，尽管我已经竭力使本书和相关配套资源接近完美，但仍然可能存在漏洞和瑕疵。欢迎读者提供关于本书的反馈意见，这将有助于我们改进和提高，以帮助更多的读者。如果你对本书有任何评论和建议，或者遇到问题需要帮助，可以致信作者的邮箱 flylolo@126.com 或本书编辑的邮箱 zhangshuang@ptpress.com.cn。

致谢

感谢出版社的编辑们对本书的耐心审阅！感谢我的家人在本书编写过程中对我的大力支持！感谢博客园中提供宝贵意见的朋友们以及我遇到的众多良师益友！

资源与支持

本书由异步社区出品，社区（https://www.epubit.com/）为您提供相关资源和后续服务。

配套资源

本书提供配套源代码，请在异步社区本书页面中点击 配套资源 ，跳转到下载界面，按提示进行操作即可。注意：为保证购书读者的权益，该操作会给出相关提示，要求输入提取码进行验证。

提交勘误

作者和编辑尽最大努力来确保书中内容的准确性，但难免会存在疏漏。欢迎您将发现的问题反馈给我们，帮助我们提升图书的质量。

当您发现错误时，请登录异步社区，按书名搜索，进入本书页面，单击“提交勘误”，输入勘误信息，点击“提交”按钮即可。本书的作者和编辑会对您提交的勘误进行审核，确认并接受后，您将获赠异步社区的100积分。积分可用于在异步社区兑换优惠券、样书或奖品。

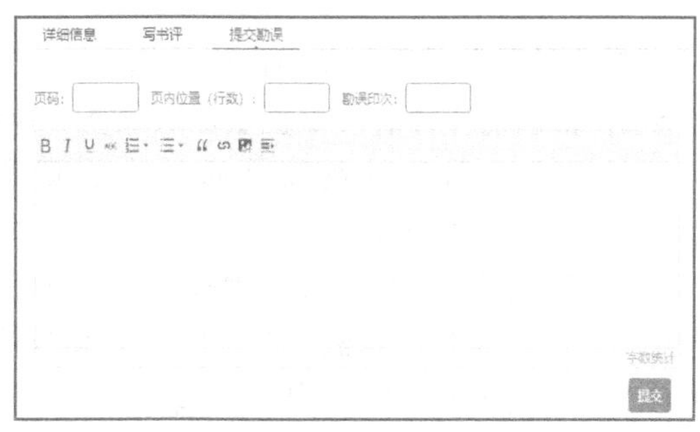

扫码关注本书

扫描下方二维码，您将会在异步社区微信服务号中看到本书信息及相关的服务提示。

与我们联系

我们的联系邮箱是 contact@epubit.com.cn。

如果您对本书有任何疑问或建议，请您发邮件给我们，并请在邮件标题中注明本书书名，以便我们更高效地做出反馈。

如果您有兴趣出版图书、录制教学视频，或者参与图书翻译、技术审校等工作，可以发邮件给我们；有意出版图书的作者也可以到异步社区在线投稿（直接访问 www.epubit.com/selfpublish/submission 即可）。

如果您所在的学校、培训机构或企业想批量购买本书或异步社区出版的其他图书，也可以发邮件给我们。

如果您在网上发现有针对异步社区出品图书的各种形式的盗版行为，包括对图书全部或部分内容的非授权传播，请您将怀疑有侵权行为的链接发邮件给我们。您的这一举动是对作者权益的保护，也是我们持续为您提供有价值的内容的动力之源。

关于异步社区和异步图书

“**异步社区**”是人民邮电出版社旗下 IT 专业图书社区，致力于出版精品 IT 技术图书和相关学习产品，为作译者提供优质出版服务。异步社区创办于 2015 年 8 月，提供大量精品 IT 技术图书和电子书，以及高品质技术文章和视频课程。更多详情请访问异步社区官网 https://www.epubit.com。

“**异步图书**”是由异步社区编辑团队策划出版的精品 IT 专业图书的品牌，依托于人民邮电出版社近 30 年的计算机图书出版积累和专业编辑团队，相关图书在封面上印有异步图书的 LOGO。异步图书的出版领域包括软件开发、大数据、AI、测试、前端、网络技术等。

异步社区

微信服务号

目　录

第 1 章　ASP.NET Core 介绍

.NET Framework 的 ASP.NET MVC 已经非常成熟稳定，为什么又会出现 ASP.NET Core 呢？它有哪些优点？它为什么要跨平台呢？通过本章的学习，读者将会得到上述问题的答案。

1.1 为什么要使用 ASP.NET Core

关于 ASP.NET Core，官方是这样定义的：ASP.NET Core 是重新设计的 ASP.NET，更改了体系结构，形成了更精简的模块化框架。ASP.NET Core 具有如下优点。

- 生成 Web UI 和 Web API 的统一场景。
- 针对可测试性进行构建。
- Razor Pages 可以使基于页面的编码方式更简单、高效。
- 能够在 Windows、macOS（也作 OSX）和 Linux 等操作系统上进行开发和运行。
- 开放源代码和以社区为中心。
- 集成新式客户端框架和开发工作流。
- 基于环境的云就绪配置系统。
- 内置依赖注入。
- 轻型的高性能模块化 HTTP 请求管道。
- 能够在 IIS、NGINX、Apache、Docker 上进行托管或在自己的进程中进行自托管。
- 可以按应用程序的依赖选择并行的.NET 版本。
- 简化新式 Web 开发的工具。

ASP.NET Core 完全作为 NuGet 包的一部分，这样便可以将应用优化为只包含必需 NuGet 包。所以与.NET Framework 版的 ASP.NET MVC 相比，ASP.NET Core 模块化更高、性能更强，加上它的跨平台特性，给了我们更多选择它的理由。

1.2 如何跨平台

.NET Framework 和 ASP.NET Core 的大体结构如图 1-1 所示。

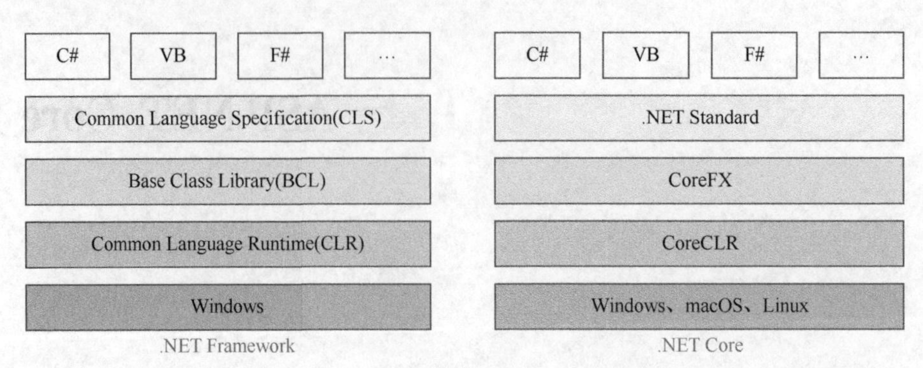

▲图 1-1

.NET Framework 本身是一个跨平台的解决方案，在这一基础上，它还支持 C#和 VB 等编程语言，且这些语言共同遵守公共语言规范（Common Language Specification，CLS），最终生成的应用程序都会被编译成公共中间语言（Common Intermediate Language，CIL）执行。从多层架构设计的角度来看，如果想让它不仅可以在多种 Windows 操作系统中运行，也可以在 macOS、Linux 操作系统中运行，则主要需要修改架构底层关于操作系统的部分，而原有的多种编程语言实际上与所运行的操作系统关系很小。

对比.NET Framework 的公共语言运行时（Common Language Runtime，CLR），ASP.NET Core 重新设计了 CoreCLR，以及一个被称为 CoreFX 的基础类库（Base Class Library，BCL）。

为了表示.NET Core 和.NET Framework 的对比关系，图 1-2 中展示了.NET Standard。这其实有些不妥，因为.NET Standard 不是包含在 ASP.NET Core 中的，它是一组 API 规范。.NET Core、.NET Framework 和 Mono 对.NET Standard 的支持关系如图 1-2 所示（N/A 表示暂不支持）。

.NET Standard	1.0	1.1	1.2	1.3	1.4	1.5	1.6	2.0	2.1
.NET Core	1.0	1.0	1.0	1.0	1.0	1.0	1.0	2.0	3.0
.NET Framework	4.5	4.5	4.5.1	4.6	4.6.1	4.6.1	4.6.1	4.6.1	N/A
Mono	4.6	4.6	4.6	4.6	4.6	4.6	4.6	5.4	6.4

▲图 1-2

引用微软官方帮助文档中的一幅图，如图 1-3 所示。图中通过饼图和柱状图两种形式展示了 CoreFX 中专门针对各个操作系统（Windows、Unix、Linux 和 OSX）的 C#代码比例和代码行数，可以看到 90%的 CoreFX 代码是与操作系统无关的，所以我们在使用 ASP.NET Core 时受不同操作系统的影响也同样会非常小。

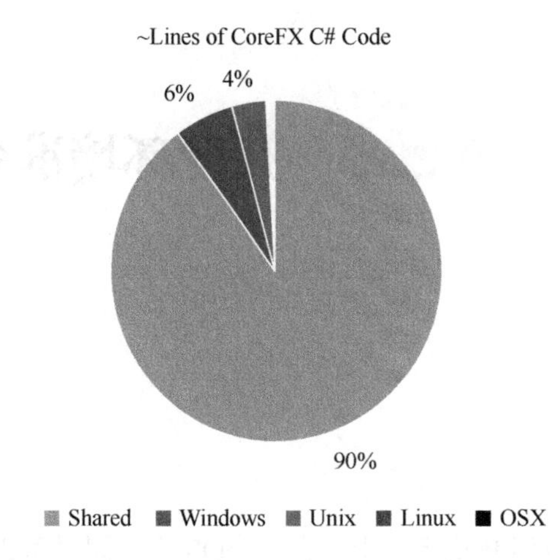
~Lines of CoreFX C# Code
6%
4%
90%
Shared
Windows
Unix
Linux
OSX

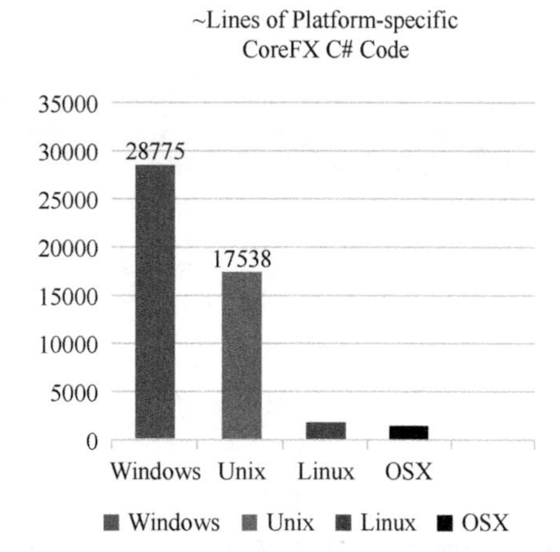
~Lines of Platform-specific CoreFX C# Code
35000
30000
25000
20000
15000
10000
5000
0
28775
17538
Windows
Unix
Linux
OSX
Windows
Unix
Linux
OSX

▲图 1-3

第2章　开发环境准备

ASP.NET Core 是跨平台的，这不止体现在应用部署及运行上，在 macOS 中，同样可以通过 Visual Studio 进行开发。本章将简要介绍开发环境的准备，这里不是通过一张张的步骤截图进行介绍，而是只对关键部分加以说明。

2.1 概述

作为.NET Core 的开发环境，.NET Core 3.0 支持 Visual Studio 2019 版本 16.3 以上，本书采用的版本是 Visual Studio Community 2019 为例。2019 年 4 月微软发布了 Visual Studio 2019，除了一些新增的功能外，其他与 Visual Studio 2017 相差不太多。

Visual Studio 下载地址见微软官网，读者可以根据自己的实际情况选择 Windows 版本或 macOS 版本。正如 90%的 CoreFX 代码都与操作系统无关一样，本书中涉及的内容也基本与 Visual Studio 的版本和操作系统无关，所以如果读者使用的 Visual Studio 版本或操作系统与本书使用的不一致，影响也不大。由于安装方式比较简单，这里就不逐步演示了，下面主要介绍需要安装的组件及说明。

2.2 Windows 环境下 Visual Studio 的安装

环境要求：Windows 7 SP1 以上版本。

下载安装的过程中会出现安装组件的选择页面，本书只涉及 ASP.NET Core 的开发，可只选择“ASP.NET 和 Web 开发”和“.NET Core 跨平台开发”，如图 2-1 所示。

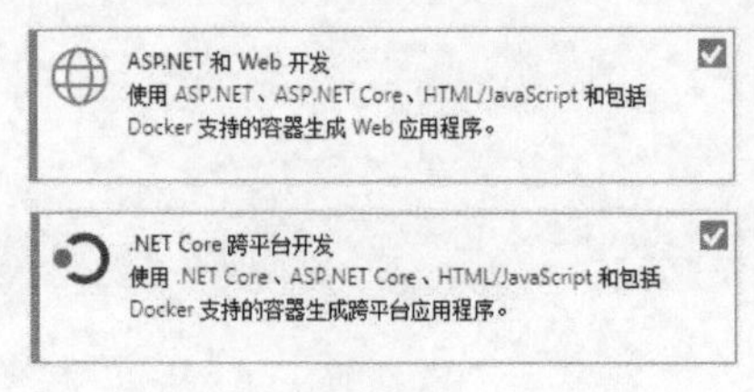

▲图 2-1

2.3 macOS 环境下 Visual Studio 的安装

自 Visual Studio 2017 以来，Visual Studio 增加了 macOS 版本，功能较 Visual Studio 的 Windows 版本要少，但目前也能满足常见的开发需求。

环境要求：macOS Sierra 10.12 及以上版本。

选择安装组件，默认如图 2-2 所示。

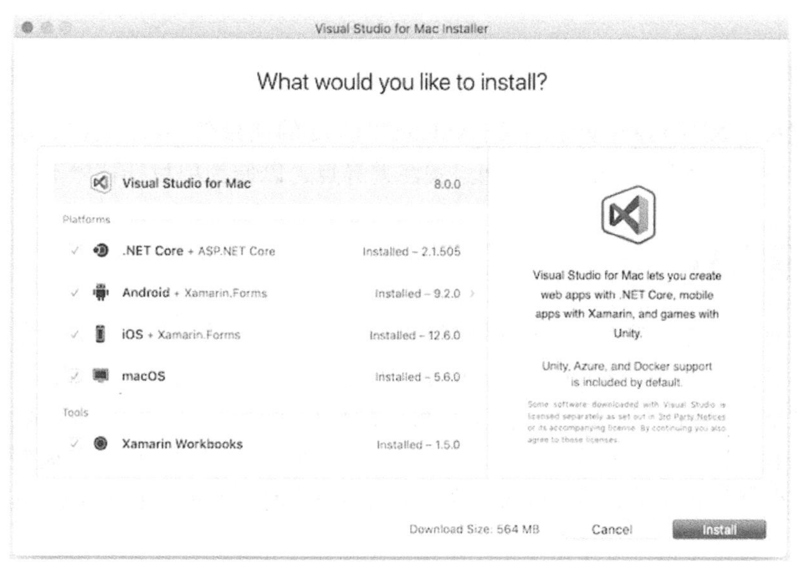

▲图 2-2

- **.NET Core**：用于 ASP.NET Core 开发。
- **Android**：用于 Android 开发，会安装 Java SDK 和 Android SDK，实际安装的内容比较多。
- **iOS**：用于 iOS 开发，需要 Xcode。
- **macOS**：用于 macOS 开发，需要 Xcode。

例如本书只涉及 ASP.NET Core 开发，只选择第一项即可。选择好安装的组件之后，单击“Install”按钮，然后等待安装完成即可。如果只安装.NET Core，那么实际安装的内容很少，很快就能安装完毕。

第3章 项目结构

本章通过对比.NET Framework 介绍 ASP.NET Core 的项目结构（下文也尽量通过这种方式进行对比，以方便读者学习和理解），通过关注差异点，以便为项目迁移做准备。

3.1 新建项目

新建项目时，首先选择 ASP.NET Core Web 应用程序，然后单击“下一步”按钮，在新窗口中输入项目名称，例如“HelloCore”。单击“创建”按钮，此时会出现如图 3-1 所示的对话框。其中有“空”“API”“Web 应用程序”和“Web 应用程序（模型视图控制器）”等几种类型可供选择。我们选择“Web 应用程序（模型视图控制器）”，没有标注“（模型视图控制器）”的“Web 应用程序”是包含示例 ASP.NET Core Razor 页面内容的 ASP.NET Core 应用程序的项目模板。

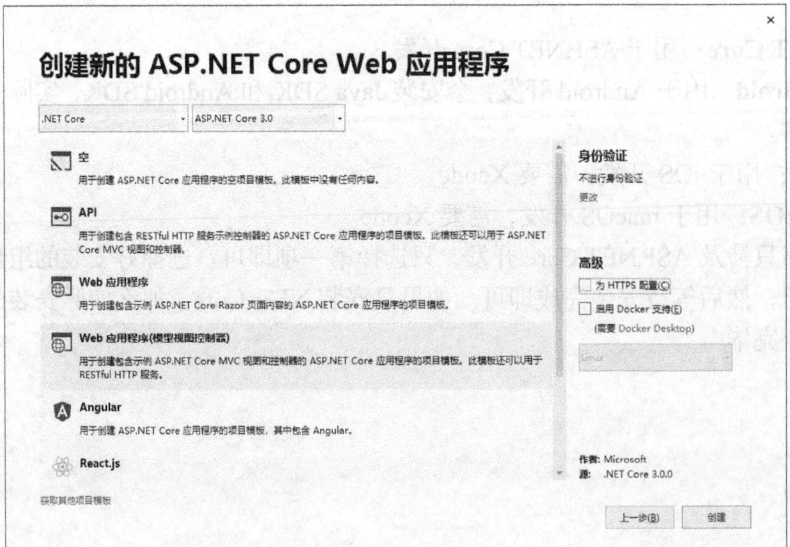

▲图 3-1

3.2 项目结构详解

新建的项目结构如图 3-2 所示。

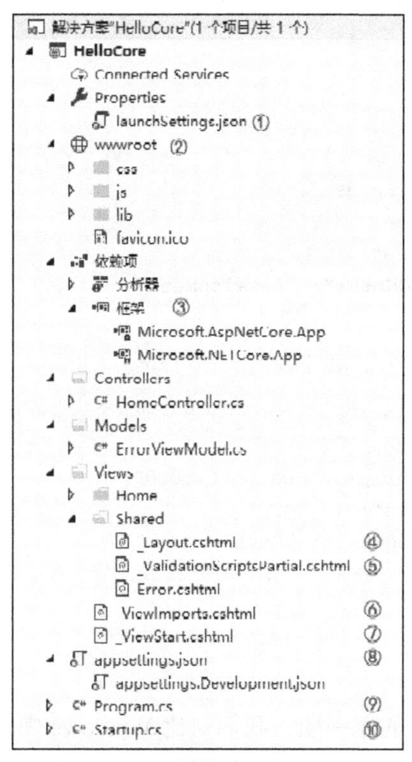

▲图 3-2

ASP.NET Core Web 应用程序的项目结构大体上和.NET Framework 一致。本章先简单介绍这些目录或文件的用途，下文再对其进行详细说明。

注意，Controller 和 Model 在此处不作介绍。对于 View，这里也只是介绍其中几个特殊的 View。

下面对照图 3-2 中的数字编号，逐一介绍它们的功能。

3.2.1 launchSettings.json

顾名思义，launchSettings.json 是一个启动配置文件，为 JSON 格式。通过图 3-2 所示的项目结构图可以发现，在 Framework 中常见的“web.config”“app.config”等 XML 格式的“.config”类型的文件在这里找不到了。

浏览 launchSettings.json，默认情况下的内容如下所示。

```
{
  "iisSettings": { //对应图 3-3
    "windowsAuthentication": false,
    "anonymousAuthentication": true,
    "iisExpress": {
      "applicationUrl": "http://localhost:65425",
      "sslPort": 0
    }
  },
  "profiles": {  //对应图 3-4
    "IIS Express": {
      "commandName": "IISExpress",
      "launchBrowser": true,
      "environmentVariables": {
        "ASPNETCORE_ENVIRONMENT": "Development"
      }
    },
    "HelloCore": {
      "commandName": "Project",
      "launchBrowser": true,
      "applicationUrl": "http://localhost:5000",
      "environmentVariables": {
        "ASPNETCORE_ENVIRONMENT": "Development"
      }
    }
  }
}
```

为了便于理解上述代码中的每一项，我们对比 Windows 操作系统中的 Visual Studio 2019 的图形化配置页面（右击当前项目，然后选择“属性”→“调试”）就容易理解了。

launchSettings.json 包含 profiles 和 iisSettings 两个对象，其中的 iisSettings 对应图 3-3，涉及我们比较熟悉的 URL、身份验证及 SSL 等配置。

Web 服务器设置

应用 URL(P): http://localhost:65425

IIS Express 位数(B): 默认值

托管模型(M): 默认

☐ 启用 SSL(S)

☑ 启用匿名身份验证(Y)

☐ 启用 Windows 身份验证(W)

▲图 3-3

而在这个图形化配置页面的最上方，有一个“配置文件”下拉列表框，其中的选项恰好对应了 profiles 中的 IIS Express 和 HelloCore，如图 3-4 所示。切换这两个选项，可以看到页面的内容也会随之改变，相当于两个页面，每个页面中的配置对应 JSON 格式文件中的 IIS Express

和 HelloCore 两个相应的节点。

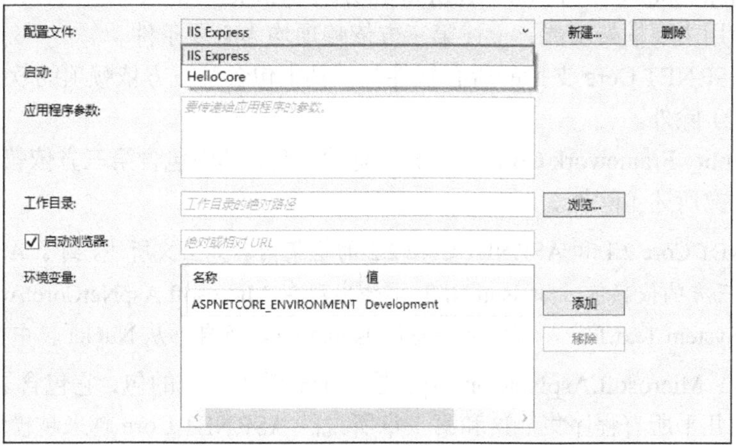

▲图 3-4

3.2.2　wwwroot

wwwroot 这个目录似乎是 IIS 的默认网站根目录，但在这里，它仅包含所有的“前端”的静态文件，包括两个常见的名为“css”和“js”的文件夹，及一个名为“lib”的文件夹，系统默认引用的 Bootstrap 和 jQuery，相关文件被放在 lib 文件夹中。

wwwroot 目录用于存放静态文件，访问其中的文件默认不需要授权，也不必使用 Controller/Action 访问路径，也就是说，可以像访问静态文件一样访问它。例如，要访问 wwwroot 目录下的文件“1.jpg”，直接访问 http://localhost:65425/1.jpg 即可。

在 Startup 文件中，会调用一个无参数的方法 UseStaticFiles()，将 wwwroot 目录标记到网站根目录。

```
public void Configure(IApplicationBuilder app, IHostingEnvironment env)
{
    //……
    app.UseStaticFiles();
    //……
}
```

具体静态文件的路径、相关自定义配置，以及授权等将在后续章节详细描述。

3.2.3　框架

这里主要分两部分：Microsoft.AspNetCore.App 和 Microsoft.NETCore.App。它们是两个比较大的程序集。

Microsoft.AspNetCore.App：它是一个内容非常多的软件包，其中包含 MVC、Razor、EF 等程序集。

微软官方对于它包含的内容的解释如下。

- 不包括除 Json.NET、Remotion.Linq 和 IX-Async 之外的其他第三方依赖项。为了确保正常使用主要框架功能，上述第三方依赖项均为必要条件。
- 包括 ASP.NET Core 支持的所有软件包，其中包含第三方依赖项的软件包（不包括上文所述）除外。
- 包括 Entity Framework Core 支持的所有软件包，其中包含第三方依赖项的软件包（不包括上文所述）除外。

注意：ASP.NET Core 2.1 和 ASP.NET Core 2.2 的框架内容如上文所述，到了 ASP.NET Core 3.0，它逐步去掉了第三方的依赖包，如 Json.NET 不会出现在 Microsoft.AspNetCore.App 中，会默认采用新的 API —— System.Text.Json，若仍需要使用 Json.NET，可自行从 NuGet 包中查找并添加。

从内容上看，Microsoft.AspNetCore.App 是一个非常大而全的包，它包含了 ASP.NET Core 中可能会用到的几乎所有程序集。这和第 1 章所说的 ASP.NET Core 越来越模块化的思想有点不一致，而且这个包会无缘无故地让项目引用一些根本用不到的程序集。事实上，它包含的这些程序集不会随着项目一起出现在部署包中，不只没有出现被项目引用的程序集，就连已经被项目引用的也不会出现。这些程序集已经存在于部署环境中，发布包不会变大，反而会变小，因此不必担心。

Microsoft.NETCore.App：它是.NET Core 的部分库，也就是.NETCore.App 框架，它依赖于更基础的 NETStandard.Library。相对于 Microsoft.AspNetCore.App，Microsoft.NETCore.App 同样包含一些程序集，但它更"基础"一些。

介绍了 Microsoft.AspNetCore.App 和 Microsoft.NETCore.App 两个包后，接下来介绍它们的区别。Microsoft.AspNetCore.App 中大部分是以"Microsoft."开头的程序集，而 Microsoft.NETCore.App 中大多数程序集是我们熟悉的"system.*XXX*"类型。二者的关系就像 ASP.NET 相对于.NET，此处是 ASP.NET Core 相对于.NET Core。

Microsoft.NETCore.App 同样是一个大而全的程序集包。在部署时，其中的程序集无论是否被引用都不会出现在部署包中，这样会显得比较精简，如图 3-5 所示。

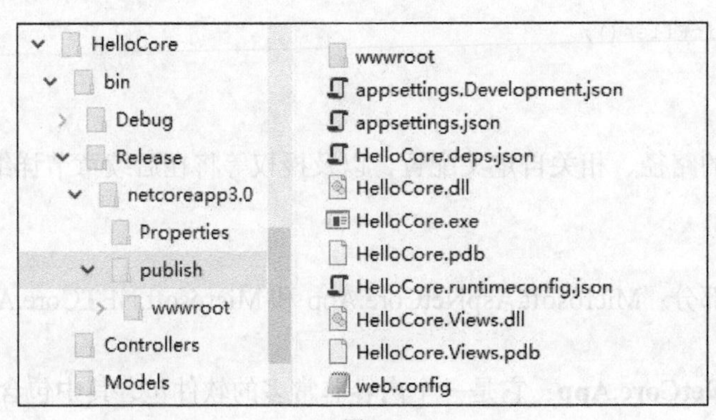

▲图 3-5

3.2.4 _Layout.cshtml

_Layout.cshtml 是一个默认的布局模板，简单来说，就是所有采用此模板的页面拥有大体一致的布局。例如，我们经常浏览的页面很多采用的是如图 3-6 所示的结构。

- **Header**：用于放置网站名称、顶部导航等信息。
- **Navigation**：左侧导航条。
- **Content**：页面的主要内容区域。
- **Footer**：放置版权、备案，以及其他一些固定在页面底部的内容。

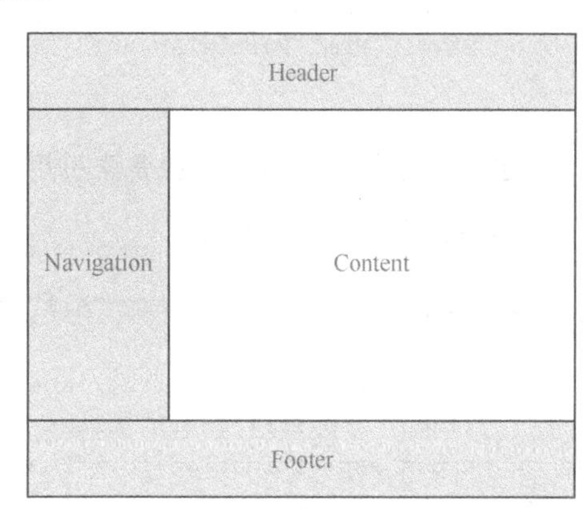

▲图 3-6

其中，Header、Footer 和 Navigation 部分基本上是不变的，变化的只是 Content 部分。打开_Layout.cshtml，我们可以看到一个@RenderBody()标识，它其实就是用来定义 Content 部分的，继承此模板的页面只需要提供这部分内容。

在_Layout.cshtml 文件中，还有类似@RenderSection("Scripts", required: false)这样的标识，引用此模板的页面可以将该页面的特定的 JavaScript 引用放在对应的 Section 中。

引用此模板，只需在代码文件的起始部分进行如下配置：

```
@{
    Layout = "~/Views/Shared/_Layout.cshtml";
}
```

如果每个页面都这样配置会比较麻烦，下文将要提到的_ViewStart.cshtml 可以解决这个问题。

3.2.5 _ValidationScriptsPartial.cshtml

_ValidationScriptsPartial.cshtml 可以用来进行验证。打开此页面，可以看到一些类似下面代码的引用方式，这段代码就是引用了 jQuery 的两个用于数据验证相关的文件，如下：

```
<environment include="Development">
   <script src="~/lib/jquery-validation/dist/jquery.validate.js"></script>
   <script src="~/lib/jquery-validation-unobtrusive/jquery.validate.unobtrusive.js">
</script>
</environment>
```

我们经常遇到如图 3-7 所示的页面信息。

▲图 3-7

当输入的格式不正确时，系统会给出提示信息。我们经常是在输入后或者提交前用 JavaScript 将输入的内容通过正则表达式验证。有了_ValidationScriptsPartial.cshtml，就不用那么麻烦了，可以通过如下方式实现。

首先，需要在相应页面通过如下代码引用_ValidationScriptsPartial.cshtml：

```
@section Scripts {
    @await Html.PartialAsync("_ValidationScriptsPartial")
}
```

注意：默认的_Layout 模板是未引用的，因为不是所有页面都需要有输入及验证操作。

然后，在页面添加验证：

```
<div asp-validation-summary="All" class="text-danger"></div>
<div class="form-group">
    <label asp-for="Email"></label>
    <input asp-for="Email" class="form-control" />
    <span asp-validation-for="Email" class="text-danger"></span>
</div>
```

最后，需要在 Model 中设置验证规则：

```
        [Required(ErrorMessage ="用户名不能为空!")]
        [Display(Name = "用户名")]
        public string UserName { get; set; }

        [EmailAddress(ErrorMessage ="Email 格式不正确!")]
        [Required]
        [Display(Name = "Email")]
        public string EMail { get; set; }
```

这样就实现了如图 3-7 所示的输入验证功能。

3.2.6　_ViewImports.cshtml

在.NET Framework 中，项目的 Views 目录下有一个 Web.config 文件，如图 3-8 所示。

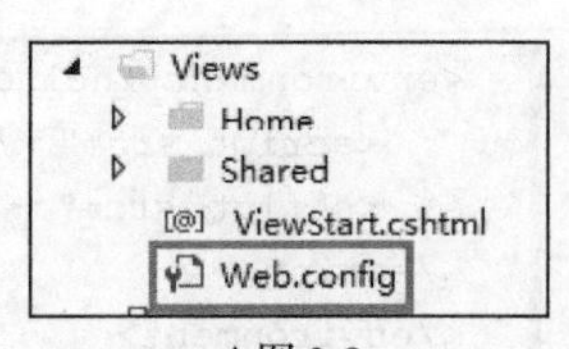

▲图 3-8

在 Views 中引用 Model 时，为了避免写“using...”，我们可以在这个 Web.config 文件中通过如下代码添加这些引用：

```
  <system.web.webPages.razor>
    <host factoryType="System.Web.Mvc.MvcWebRazorHostFactory, System.Web.Mvc, Version
=5.2.3.0, Culture=neutral, PublicKeyToken=31BF3856AD364E35" />
    <pages pageBaseType="System.Web.Mvc.WebViewPage">
      <namespaces>
        <add namespace="System.Web.Mvc" />
```

```
        <add namespace="System.Web.Mvc.Ajax" />
        <add namespace="System.Web.Mvc.Html" />
        <add namespace="System.Web.Optimization"/>
        <add namespace="System.Web.Routing" />
        <add namespace="HelloWorldDemo" />
        <add namespace="HelloWorldDemo.Model" />
      </namespaces>
    </pages>
  </system.web.webPages.razor>
```

现在这个 Web.config 文件被替换成了接下来要讲的_ViewImports.cshtml，打开_ViewImports.cshtml 文件可以看到如下代码：

```
@using HelloWorldDemo
@using HelloWorldDemo.Models
@addTagHelper *, Microsoft.AspNetCore.Mvc.TagHelpers
```

替换成_ViewImports.cshtml 其实就是为了实现 Web.config 文件的功能，但它又比 Web.config 文件简洁很多。这里有一个比较特别的地方，_ViewImports.cshtml 文件中出现了一个@addTagHelper，它的作用是什么呢？在上文提到的验证中，我们看到过这样的代码：

```
    <label asp-for="Email"></label>
    <input asp-for="Email" class="form-control" />
    <span asp-validation-for="Email" class="text-danger"></span>
```

而在 ASP.NET MVC 的 Framework 版本中，相应的代码如下：

```
@Html.LabelFor(m => m.EMail)
@Html.EditorFor(m => m.EMail)
@Html.ValidationMessageFor(m=>m.EMail)
```

从表面上看，TagHelper 和 HtmlHelper 的功能有点像，只是 TagHelper 的书写方式更像 HTML，这使前端工程师不至于看到"HTML"中混杂着@Html.EditorFor 就觉得"另类"。关于 TagHelper 先介绍这些，后续章节会有详细讲解。

3.2.7 _ViewStart.cshtml

打开_ViewStart.cshtml 文件会看到如下代码：

```
@{
    Layout = "_Layout";
}
```

_ViewStart.cshtml 文件中的内容会在所有 View 被执行前执行，上述这行代码就是给所有的 View 一个默认的 Layout 模板。因此，在一个名为 Index.cshtml 的 View 中写成这样：

```
{
    Layout = null;
}
```

或写成这样：

```
@{

}
```

这两种写法是不一样的。第一种写法是设置 Index.cshtml 不采用任何模板；第二种写法是什么都没做，它会采用_ViewStart.cshtml 中指定的模板。

当然，_ViewStart.cshtml 文件的作用不只是写这么一句代码，我们还可以在这里写一些其他需要“通用”执行的代码。

3.2.8　appsettings.json 和 appsettings.Development.json

appsettings.json 和 appsettings.Development.json 这两个文件就是原来的 ASP.NET MVC 的 Framework 版本中项目根目录下的 Web.config 文件，默认情况只有几行代码，非常精简，且只有 log（日志）的相关配置。在正常项目中，我们要配置的肯定不止这一点。举个例子，添加数据库连接配置：

```
{
  "ConnectionStrings": {
    "DefaultConnection":
"Server=(localdb)\\mssqllocaldb;Database=ContosoUniversity1;Trusted_Connection=True;
MultipleActiveResultSets=true"
  },
  "Logging": {
    "IncludeScopes": false,
    "LogLevel": {
      "Default": "Warning"
    }
  }
}
```

3.2.9　Program.cs

打开 Program.cs 文件，可以看到它包含一个我们非常熟悉的 Main 方法，也就是应用的起点。项目启动后，通过 UseStartup<Startup>()指定下文的 Startup 文件进行启动配置，代码如下：

```
public class Program
{
    public static void Main(string[] args)
    {
```

```
            CreateHostBuilder(args).Build().Run();
        }

        public static IHostBuilder CreateHostBuilder(string[] args) =>
            Host.CreateDefaultBuilder(args)
                .ConfigureWebHostDefaults(webBuilder =>
                {
                    webBuilder.UseStartup<Startup>();
                });
    }
```

3.2.10 Startup.cs

Startup.cs 是 ASP.NET Core 中非常重要的文件，加载配置、通过依赖注入加载组件和注册路由等都在此处进行。下列为其默认的代码：

```
public class Startup
{
    public Startup(IConfiguration configuration)
    {
        Configuration = configuration;
    }

    public IConfiguration Configuration { get; }

    public void ConfigureServices(IServiceCollection services)
    {
        services.AddControllersWithViews();
    }
    public void Configure(IApplicationBuilder app, IWebHostEnvironment env)
    {
        if (env.IsDevelopment())
        {
            app.UseDeveloperExceptionPage();
        }
        else
        {
            app.UseExceptionHandler("/Home/Error");
        }
        app.UseStaticFiles();

        app.UseRouting();

        app.UseAuthorization();

        app.UseEndpoints(endpoints =>
```

```
            {
                endpoints.MapControllerRoute(
                    name: "default",
                    pattern: "{controller=Home}/{action=Index}/{id?}");
            });
        }
    }
```

ConfigureServices 方法主要用于各种服务配置的依赖注入，而 Configure 方法主要用于管道配置。在这里，我们可以通过中间件的方式向管道中插入我们需要的内容。例如，可以通过 env.IsDevelopment 设置当前环境不同状态下的错误页，通过 app.UseStaticFiles()指定静态文件，还可以用 app.UseAuthentication()来进行身份验证等。

在这里可以使用 Use、Run 和 Map 来配置 HTTP 管道。

- Use 方法可使管道短路（即不调用 Next 方法请求委托）。
- Run 是一种约定，并且某些中间件组件可公开在管道末尾运行的 Run[Middleware]方法。
- Map 用于创建管道分支。

此处涉及的内容非常多，例如管道机制、路由注册、身份认证等，都需要进行专题讲解，我们会在下文进行详细描述。

第 4 章　_Layout 与_ViewStart

本章我们新建一个项目，并通过这个项目熟悉一下_Layout 与_ViewStart 以及它们的加载顺序。

4.1 _Layout 的应用

新建项目的默认运行效果如图 4-1 所示。

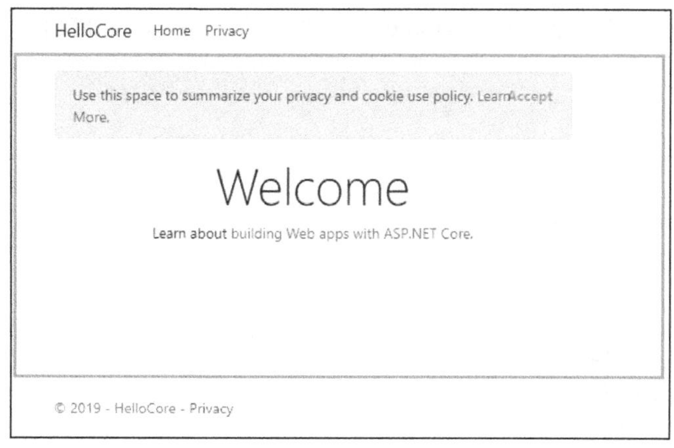

▲图 4-1

页面被图中的方框分为上、中、下 3 部分：上面的菜单栏、中间的内容部分，以及下面的版权区域。单击菜单栏中的“Home”或“Privacy”切换页面，可以看到上下两部分都是不变的，只有中间部分的内容在变。

打开 Shared 文件夹中的_layout.cshtml 文件，它的 body 标签内容结构如下：

```
<body>
    <header>
       //菜单栏代码，已省略
    </header>
```

```
    <div class="container">
        <partial name="_CookieConsentPartial" />
        <main role="main" class="pb-3">
            @RenderBody()
        </main>
    </div>

    <footer class="border-top footer text-muted">
        <div class="container">
            &copy; 2019 - HelloCore - <a asp-area="" asp-controller="Home" asp-action
="Privacy">Privacy</a>
        </div>
</footer>
//JavaScript 文件引用代码，已省略

</body>
```

与页面的上、中、下结构对应，这里有 header、container、footer 3 个标签。而 container 标签中的<partial name="_CookieConsentPartial" />是与隐私条款相关的，单击“Accept”按钮接受条款，该条目消失，所以@RenderBody()就是页面中间变化的部分了。

现在将主页内容改变一下，打开 Home 文件夹下的 Index.cshtml 文件，将里面的代码改成如下代码：

```
@{
    ViewData["Title"] = "主页";
}
<table class="table table-hover">
    <thead>
        <tr>
            <th>序号</th>
            <th>日期</th>
            <th>体重</th>
            <th>备注</th>
        </tr>
    </thead>
    <tbody>
        <tr>
            <td>1</td>
            <td>2018-02-15</td>
            <td>66.6</td>
            <td>除夕,胖了</td>
        </tr>
        <tr>
            <td>2</td>
```

```
                <td>2018-02-16</td>
                <td>68.8</td>
                <td>春节,又重了</td>
            </tr>
        </tbody>
    </table>
```

刷新页面，显示结果如图 4-2 所示。

HelloCore Home Privacy

序号	日期	体重	备注
1	2018-02-15	66.6	除夕,胖了
2	2018-02-16	68.8	春节,又重了

© 2019 - HelloCore - Privacy

▲图 4-2

看起来效果还不错。读者可能会注意到，这个 table 标签有一个名为“table table-hover”的 class，这个 class 定义在哪里呢?

再次打开_layout.cshtml 文件，可以看到在 head 标签中通过如下代码引用了 Bootstrap 的 css：

```
    <environment include="Development">
        <link rel="stylesheet" href="~/lib/bootstrap/dist/css/bootstrap.css" />
    </environment>
```

所以可以把一些“通用”的 css 和 js 的引用放在_layout.cshtml 文件里，避免重复写这些引用。顺便在 footer 标签中添加“--改变 Layout 文件”，然后分别单击菜单中的“Home”和“Privacy”查看相应页面，发现它们的 Footer 都添加了“--改变 Layout 文件”这句话，效果如图 4-3 所示。

▲图 4-3

看到这里读者可能会有疑问：对应的 Index.cshtml 和 Privacy.cshtml 文件中都没有对 Layout 做引用，那么是通过什么方式引用的呢？继续往下看会找到答案。

4.2 _ViewStart 的应用

回顾修改后的 Index.cshtml 页面的代码，我们并没有在相应代码文件中写 Layout = "_Layout" 这样的代码，这是因为已经在_ViewStart.cshtml 中默认设置了。_ViewStart.cshtml 中只有如下代码：

```
@{
    Layout = "_Layout";
}
```

如果我们在 Index 页面相关代码中添加代码：

```
@{
    Layout=null;
    ViewData["Title"] = "主页";
}
```

再次刷新页面，效果如图 4-4 所示。

序号	日期	体重	备注
1	2018-02-15	66.6	除夕，胖子
2	2018-02-16	68.8	春节，又重了

▲图 4-4

Header 和 Footer 内容全都没有了，Table 的样式也没有了。这是因为这些本来是写在_Layout.cshtml 中的，现在失去了对_Layout.cshtml 文件的引用，这些效果也就消失了。

总结：_ViewStart.cshtml 代码文件中对模板页做了默认的设置，除非显式地写明 Layout=*XXX*，否则会采用_ViewStart.cshtml 中的默认设置。所以未做设置和设置 Layout = "_Layout" 的效果是一样的。

4.3 页面的加载顺序

几个页面的加载顺序是_ViewStart.cshtml→Index.cshtml→_Layout.cshtml。

- 首先_ViewStart 在所有 View 加载之前加载，设置了默认的模板页。
- 接着由 Controller 指定的页面查找 Index.cshtml 加载，并读取该页面的 Layout 设置。
- 最后根据 Index 页面的 Layout 设置的模板页查找对应的模板页加载。

将_ViewStart 中的 Layout = "_Layout" 改为 Layout = "_Layout1"，再次运行，页面会出现如下找不到模板页的错误。

```
An unhandled exception occurred while processing the request.
InvalidOperationException: The layout view '_Layout1' could not be located. The following
locations were searched:
```

```
/Views/Home/_Layout1.cshtml
/Views/Shared/_Layout1.cshtml
/Pages/Shared/_Layout1.cshtml
```

View 的查找规则：先查找 Controller 对应的文件夹（这里是 Home），若未找到，则到 Views/Shared 和 Pages/Shared 文件夹查找；若最终未找到，则提示错误。

第5章　TagHelper

5.1 概述

TagHelper 是 ASP.NET Core 中新出现的一个名词，它的作用是使服务器端代码可以在 Razor 文件中参与创建和显示 HTML 标签。目前，在 ASP.NET Core 中，TagHelper 的功能类似于 HtmlHelper。

首先通过一个例子来看 TagHelper 是怎么使用的，它和 HtmlHelper 有什么区别。新建一个 Book 类：

```
public class Book
{
    [Display(Name = "编号")]
    public string Code { get; set; }
    [Display(Name = "名称")]
    public string Name { get; set; }
}
```

新建对应的 Controller 和 Action：

```
public class BookController : Controller
{
    public IActionResult Index()
    {
        return View(new Book() { Code = "001", Name = "ASP" });
    }
}
```

最后新建对应的 View：

```
@model Book
@{
    Layout = null;
```

```
}
@Html.LabelFor(m => m.Name)
@Html.EditorFor(m => m.Name)
<br />
<label asp-for="Name"></label>
<input asp-for="Name" />
```

这里分别通过 HtmlHelper 和 TagHelper 实现了一个文本和一个输入框的显示。查看网页源代码，可以看到二者生成的 HTML 代码如下：

```
<label for="Name">Name</label>
<input class="text-box single-line" id="Name" name="Name" type="text" value="ASP" />
<br/>
<label for="Name">Name</label>
<input type="text" id="Name" name="Name" value="ASP" />
```

目前看起来二者差不多，从工作量来看区别不大。现在功能实现了，需要做一些样式处理。简单举例说明，现在希望 Book 的编号（Code）对应的 label 的颜色为红色，定义一个 CSS 如下：

```
<style type="text/css">
    .codeColor {
        color:red;
    }
</style>
```

然后准备把这个样式应用到 label 上，这时如果是 HtmlHelper，很有可能会被人问："class 写在哪？"我们可以这样回答：

```
@Html.LabelFor(m=>m.Name,new {@class="codeColor"})
```

一些前端工程师添加后达到了想要的效果，但可能记不住这种写法，下次还会问同样的问题。

如果使用 TagHelper 就方便了，只需像平时给 HTML 的标签添加 class 一样操作即可，也就是：

```
<label asp-for="Name" class="codeColor"></label>
```

一些前端工程师表示这种写法"真是太友好了"。同样对于 Form 及验证，比较一下两种写法的不同。

HtmlHelper 版：

```
@using (Html.BeginForm("Index", "Home", FormMethod.Post)){
@Html.LabelFor(m => m.Code)
@Html.EditorFor(m => m.Code)
@Html.ValidationMessageFor(m => m.Code)
<input type="submit" value="提交" />
}
```

TagHelper 版：

```
<form asp-action="Index" asp-controller="Home" method="post">
    <label asp-for="Code"></label>
    <input asp-for="Code" />
    <span asp-validation-for="Code"></span>
    <input type="submit" value="提交" />
</form>
```

5.2 自定义 TagHelper

现在有这样的需求：用于显示 Book 的编号的 label 不只要添加名为 codeColor 的 CSS 样式，还要给 Book 的编号自动添加一个前缀，例如“BJ”。

对于这样的需求，可以通过一个简单的标记，由 TagHelper 自动实现。例如：

```
<label show-type="bookCode">1001</label>
```

自定义了一个属性 show-type，用于标识这个 label 的显示类别，“1001”为假定的 Book 的编号。通过这样的设置方式，如果将来需求有变化，需要对编号的显示效果做更多的修饰时，只需修改对应的 TagHelper，而页面内容部分的代码不需要做任何调整。

系统提供了一种方便的自定义 TagHelper 的方式，就是继承系统提供的 TagHelper，并重写它的 ProcessAsync 方法，例如下面的代码：

```
    public class LabelTagHelper : TagHelper
    {
        public override async Task ProcessAsync(TagHelperContext context, TagHelperOu
tput output)
        {
            if (output.Attributes.TryGetAttribute("show-type", out TagHelperAttribute
  showTypeAttribute))
            {
                if (showTypeAttribute.Value.ToString().Equals("bookCode"))
                {
                    output.Attributes.SetAttribute("class", "codeColor");

                    string content = output.Content.IsModified ?
output.Content.GetContent
() : (await output.GetChildContentAsync()).GetContent(); ;
                    output.Content.SetContent("BJ" + content);
                }
            }
        }
    }
```

首先判断 label 是否设置了 show-type="bookCode"属性，然后获取当前 label 的 Content，为其添加前缀 BJ 后作为此 label 的新内容。注意代码中 Content 的获取方式，如果它没有被修改，直接通过 output.Content.GetContent()来获取，得到的结果会是空的。

访问 Index 页面，可以看到该标签已被处理，如图 5-1 所示。

BJ1001

▲图 5-1

备注：

- 关于获取 show-type 的值，还可以有其他方式，下文会讲到。
- 从命名规范化的角度来看，建议将自定义的 TagHelper 命名为 *XXX*TagHelper 这样的格式。

5.3 TagHelper 的注册

TagHelper 自定义之后需要进行注册，否则它是不会生效的。打开_ViewImports.cshtml，其内容默认为：

```
@using TagHelperDemo
@using TagHelperDemo.Models
@addTagHelper *, Microsoft.AspNetCore.Mvc.TagHelpers
```

在最下面添加一条：

```
@addTagHelper *, TagHelperDemo
```

这一行代码就是将程序集 TagHelperDemo（即第 2 个参数）中的所有 TagHelper（第 1 个参数为“*”，表示所有）全部启用。假如还定义了一个 PasswordTagHelper，但只想添加 LabelTagHelper，可以这样写：

```
@addTagHelper TagHelperDemo.TagHelpers. LabelTagHelper, TagHelperDemo
```

如果想添加所有自定义的 TagHelper，但要去除 LabelTagHelper 呢？

可以先添加所有 TagHelper，再通过 removeTagHelper 方法去除这个 LabelTagHelper，如下代码所示：

```
@addTagHelper *, TagHelperDemo
@removeTagHelper TagHelperDemo.TagHelpers. LabelTagHelper, TagHelperDemo
```

5.4 TagHelper 的作用范围

在项目中，可能不只使用 label 标签来显示 Book 的 Code，还可能需要使用 p、span 等类型的标签。现在的需求是，无论是上述哪一种标签，都要实现添加 CSS 样式和内容前缀的功能。

现在在 index.cshtml 中新增一个 p 标签，代码如下：

```
<p show-type="bookCode">1002</p>
```

访问该页面发现 1002 未被处理，这是因为我们定义的 TagHelper 名为 LabelTagHelper，它在默认的情况下只会处理 label 标签。当然也可以做特殊设置，例如下面的代码：

```
[HtmlTargetElement("p")]
public class LabelTagHelper : TagHelper
{
    //代码省略
}
```

通过“[HtmlTargetElement("p")]”指定该 TagHelper 只能被用于 p 标签。再次访问此页面，发现 p 标签被处理了，而原来的 label 变为未被处理的样式。这说明显式指定目标标签的方式的优先级要高于默认的名称匹配的优先级。除了设置目标标签，还可以有一些其他的辅助设置：

```
[HtmlTargetElement("p", Attributes = "show-type", ParentTag = "div")]
public class LabelTagHelper : TagHelper
```

这样写程序会匹配拥有 show-type 属性的 P 标签，并且要求父标签为 div。这几个条件是“and”关系。如果还想匹配 label 标签，可以添加对 label 的设置，例如下面代码：

```
[HtmlTargetElement("p", Attributes = "show-type", ParentTag = "div")]
[HtmlTargetElement("label", Attributes = "show-type", ParentTag = "div")]
public class LabelTagHelper : TagHelper
```

这两个 HtmlTargetElement 的关系是“or”。通过这样的设置，可以极大地缩小目标标签的匹配范围。

但是这样设置之后，TagHelper 的名字再叫 LabelTagHelper 就不合适了，可以改为 BookCodeTagHelper，最终的代码如下：

```
[HtmlTargetElement("p", Attributes = "show-type", ParentTag = "div")]
[HtmlTargetElement("label", Attributes = "show-type", ParentTag = "div")]
public class BookCodeTagHelper : TagHelper
{
    public override async Task ProcessAsync(TagHelperContext context, TagHelperOutput
 output)
    {
        if (output.Attributes.TryGetAttribute("show-type", out TagHelperAttribute sho
wTypeAttribute))
        {
            if (showTypeAttribute.Value.ToString().Equals("bookCode"))
            {
                output.Attributes.SetAttribute("class", "codeColor");

                string content = output.Content.IsModified ? output.Content.GetContent() :
                                 (await output.GetChildContentAsync()).GetContent(); ;
                output.Content.SetContent("BJ" + content);
```

```
                }
            }
        }
    }
```

如果想使个别 HTML 标签屏蔽 TagHelper 的样式，可以使用“!”。例如下面的标签：

```
<!label show-type="bookCode">1001</label>
```

5.5 自定义标签

上一节最终形成了一个名为 BookCodeTagHelper 的 TagHelper，我们知道 LabelTagHelper 是可以按名称默认匹配 label 标签的，那么是否可以自定义一个 BookCode 标签呢？在 Index.cshtml 中添加这样的代码：

```
<BookCode>1003</BookCode>
```

由于自定义 BookCode 标签的目的就是专门显示 Book 的编号，所以不必添加 show-type 属性。修改 BookCodeTagHelper，代码如下：

```
public class BookCodeTagHelper : TagHelper
{
    public override async Task ProcessAsync(TagHelperContext context, TagHelperOutput
  output)
    {
        output.Attributes.SetAttribute("class", "codeColor");

        string content = output.Content.IsModified ? output.Content.GetContent() :
                        (await output.GetChildContentAsync()).GetContent(); ;
        output.Content.SetContent("BJ" + content);
    }
}
```

去掉了两个 HtmlTargetElement 设置以及 show-type 属性，访问 Index 页面查看新建的 BookCode 标签是否会被处理。结果是没有被处理。这是为什么呢？

这是由于 TagHelper 会将采用 Pascal 大小写格式的类和属性名转换为各自对应的含“-”格式，即将 BookCode 转换为 book-code。获取标签的属性值同样遵循这样的规则，所以将标签改为如下写法即可：

```
<book-code>1003</book-code>
```

再次运行程序进行测试，发现这个新标签被成功处理。查看网页源代码，被处理后的 HTML 代码如下：

```
<book-code class="codeColor">TJ1003</book-code>
```

如果想将其改为 label，可以在 BookCodeTagHelper 中通过指定 TagName 实现：

```
public class BookCodeTagHelper : TagHelper
{
    public Book Book { get; set; }
    public override async Task ProcessAsync(TagHelperContext context, TagHelperOutput
  output)
    {
        output.TagName = "label";
        output.Attributes.SetAttribute("class", "codeColor");

        string content = output.Content.IsModified ? output.Content.GetContent() :
                    (await output.GetChildContentAsync()).GetContent(); ;
        output.Content.SetContent(Book.Prefix + content);
    }
}
```

5.6 TagHelper 与页面之间的数据传递

假如现在的新需求是图书编号的前缀不再固定为"BJ"，需要在标签中定义，例如：

```
<book-code prefix="SH">1003</book-code>
```

获取 prefix 的值，在前面的例子中采用的是 TryGetAttribute 方法。其实还有更简单的方法，例如通过修改 BookCodeTagHelper 获取，代码如下：

```
    public class BookCodeTagHelper : TagHelper
    {
        public string Prefix { get; set; }
        public override async Task ProcessAsync(TagHelperContext context, TagHelperOutput
output)
        {
            output.Attributes.SetAttribute("class", "codeColor");

            string content = output.Content.IsModified ? output.Content.GetContent() :
                        (await output.GetChildContentAsync()).GetContent(); ;
            output.Content.SetContent(Prefix + content);
        }
    }
```

标签中 prefix 的值会自动赋值给 BookCodeTagHelper. Prefix。那么如果是 Model 中的值呢？假如 Book 有一个属性 public string Prefix { get; set; }，这和传入一个字符串没什么区别，那么可以这样写：

```
<book-code prefix="@Model.Prefix">1003</book-code>
```

这种传值方式不只支持字符串，还支持 Model 整体传入。修改标签如下：

```
<book-code book="@Model">1003</book-code>
```

修改 BookCodeTagHelper 代码：

```
    public class BookCodeTagHelper : TagHelper
    {
        public Book Book { get; set; }
        public override async Task ProcessAsync(TagHelperContext context, TagHelperOutput
output)
        {
            output.Attributes.SetAttribute("class", "codeColor");

            string content = output.Content.IsModified ? output.Content.GetContent() :
                            (await output.GetChildContentAsync()).GetContent(); ;
            output.Content.SetContent(Book.Prefix + content);
        }
    }
```

5.7 取消标签输出

前面的几个例子都是对满足条件的标签进行修改，TagHelper 也可以取消对应标签的输出，例如存在这样一个标签：

```
<div simple-type="Simple1"></div>
```

如果不想让它显示在生成的 HTML 页面中，可以这样处理：

```
[HtmlTargetElement("div",Attributes = "simple-type")]
public class Simple1TagHelpers : TagHelper
{
    public string SimpleType { get; set; }
    public override void Process(TagHelperContext context, TagHelperOutput output)
    {
        if (SimpleType.Equals("Simple1"))  //可以是其他一些判断规则
        {
            output.SuppressOutput();
        }
    }
}
```

5.8 TagBuilder

在 TagHelper 中，可以用 TagBuilder 来辅助生成标签，例如存在以下两个 div：

```
<div simple-type="Simple2"></div>
<div simple-type="Simple3"></div>
```

想在 div 中添加 HTML 标签，可以这样写：

```
[HtmlTargetElement("div",Attributes = "simple-type")]
public class Simple1TagHelpers : TagHelper
{
    public string SimpleType { get; set; }
    public override void Process(TagHelperContext context, TagHelperOutput output)
    {
        if (SimpleType.Equals("Simple2"))
        {
            output.Content.SetHtmlContent("<p>Simple2</p>");
        }
        else if (SimpleType.Equals("Simple3"))
        {
            var p = new TagBuilder("p");
            p.InnerHtml.Append("Simple3");
            output.Content.SetHtmlContent(p);
        }
    }
}
```

通过 TagBuilder 生成了一个新的 p 标签，并将它插入 div。

第6章 应用的跨平台部署

既然.NET Core可以跨平台，那么本章就介绍一下如何将网站部署到CentOS上。本章涉及的内容：CentOS下Nginx和ASP. NET Core环境的搭建、多网站多域名的配置部署、SSL证书配置、强制HTTPS访问配置、ASP. NET Core反向代理等。

6.1 概述

本章案例的服务器环境为阿里云ECS，操作系统为CentOS Linux release 7.6.1810 (Core)。我们将在服务器上搭建两个网站，功能如下。

- **网站服务**：www.*XXX*.com，用于页面访问。
- **API服务**：api. *XXX*.com，用于API服务。

注意：以上网站地址并非真实地址，只是在这里作为例子，两个网站全部强制以https的方式访问。首先在/home下新建一个名为www的文件目录，接着分别创建/home/www/web和/home/www/api两个目录用于存放上述两个网站相关文件。为了方便操作，准备如下两个工具。

- **PuTTY**：这是一个远程命令行的工具。如果是使用macOS，可以使用Termius，可以在应用商店搜到。
- **FileZilLa**：SFTP工具，用于将生成的发布包部署到CentOS。它也有macOS版本。

6.2 在CentOS中安装ASP.NET Core环境

安装ASP.NET Core环境有Runtime和SDK两种方式，区别类似于Java的JDK和JRE。Runtime只包含应用运行的基本环境，而SDK则用于开发环境，即官方提供的下载地址中使用Run Apps和Build Apps描述的这两种。因为我们不需要在CentOS上编码，所以安装Runtime就足够了。

在页面的“all downloads”区域中找到CentOS对应的Runtime进行安装，这里要注意版本问题，当前项目的版本要和本次安装的Runtime的版本一致。此处以3.0版本为例。

通过 PuTTY 将其链接到 CentOS 服务器，按照该页面上的步骤执行如下命令，此命令将会安装.NET Core Runtime 和 ASP.NET Core Runtime。

```
sudo rpm -Uvh
https://packages.microsoft.com/config/centos/7/packages-microsoft-prod.rpm
sudo yum update
sudo yum install aspnetcore-runtime-3.0
```

6.3 在 Windows 上用 Visual Studio 发布项目

分别创建两个项目，一个为 Web 项目，命名为 HelloCore；另一个为 API 项目，命名为 HelloCoreAPI。由于在默认情况下，项目采用的是 5000 端口，所以为了避免这两个项目发生端口冲突，修改 Program.cs，将端口改为常用的 8080 和 8081：

```
public class Program
  {
      //省略了 main 方法
      public static IHostBuilder CreateHostBuilder(string[] args) =>
          Host.CreateDefaultBuilder(args)
              .ConfigureWebHostDefaults(webBuilder =>
              {
                  webBuilder.UseUrls("http://*:8081").UseStartup<Startup>();
              });
  }
```

分别右击两个项目选择发布到文件夹，其他采用默认设置。默认情况下是 FDD（依赖框架部署），发布生成的内容不包含依赖的框架内容，将依赖前文安装的 Runtime。

在 CentOS 上创建两个文件夹/home/www/web 和/home/www/api，通过 FileZilla 将发布的文件上传到该文件夹（支持拖动）。默认的发布位置在项目的 bin\Release\netcoreapp3.0\publish 目录。

参考创建目录命令：mkdir -p /home/www/web。

6.4 项目运行测试

执行命令运行上传后的项目：

```
dotnet /home/www/web/HelloCore.dll
```

结果如图 6-1 所示，则表示运行成功，提示 Kestrel 开始监听 8080 端口。此时浏览器无法访问 http://服务器 IP:8081，若想测试，可以在实例的安全组中配置开放 8081 端口，并且建议测试后关闭，因为我们会通过 Nginx 访问此服务器，而不是将此端口开放允许外部访问。对 Web、API 项目也进行同样的操作。至此，配置的两个项目都可以正常访问了。

读者可能会有疑惑，既然之前的项目已经可以正常访问了，为什么还要用 NGINX？在项目中直接指定监听 80 端口不就可以了吗？因为如果这样做，该服务器会直接占用 80 端口，但像现在这样的情况，我们需要将来自不同域名的访问指定到不同的端口处理，例如将 a.com 的请求指定到 8080 端口，将 b.com 的请求指定到 8081 端口。另外，每次通过命令 dotnet *XXX*.dll 来启动也不是一个很好的方式，我们可以创建一个 service 来管理它，这有点像 Windows 操作系统的 service。

```
info: Microsoft.Hosting.Lifetime[0]
      Now listening on: http://[::]:8080
info: Microsoft.Hosting.Lifetime[0]
      Application started. Press Ctrl+C to shut down.
```

▲图 6-1

6.5 创建 service 管理应用

用 nano 创建一个 service 文件，命令如下：

```
sudo nano /etc/systemd/system/kestrel-web.service
```

nano 是一个文本编辑工具，如果提示“nano: command not found”，则表示 nano 没有安装，执行“yum install nano”命令安装即可。也可以新建文本文件，写入下面的文件内容，然后用 FileZilla 将该文件放置到对应目录：

```
[Unit]
Description=Example .NET Web App running on CentOS 7

[Service]
WorkingDirectory=/home/www/api
ExecStart=/usr/bin/dotnet /home/www/web/HelloCore.dll
Restart=always
# Restart service after 10 seconds if dotnet service crashes
RestartSec=10
SyslogIdentifier=dotnet-example
User=root
Environment=ASPNETCORE_ENVIRONMENT=Production

[Install]
WantedBy=multi-user.target
```

保存并启动服务：

```
systemctl enable kestrel-web.service
systemctl start kestrel-web.service
```

查看是否成功：

```
systemctl status kestrel-web.service
```

成功后的提示大概如图 6-2 所示。

```
systemd[1]: Started Example .NET Web App running on CentOS 7.
dotnet-example[22742]: info: Microsoft.Hosting.Lifetime[0]
dotnet-example[22742]: Now listening on: http://[::]:8080
dotnet-example[22742]: info: Microsoft.Hosting.Lifetime[0]
dotnet-example[22742]: Application started. Press Ctrl+C to shut down.
dotnet-example[22742]: info: Microsoft.Hosting.Lifetime[0]
dotnet-example[22742]: Hosting environment: Production
dotnet-example[22742]: info: Microsoft.Hosting.Lifetime[0]
dotnet-example[22742]: Content root path: /home/www/api
```

▲图 6-2

在此处若遇到问题，可以执行 journalctl -xe 查看错误信息。按照错误提示修改该文件，修改后再次执行程序，会提示先执行 systemctl daemon-reload 重新加载 service 文件。

为 API 服务做相同的操作进行配置。

6.6 安装 Nginx

安装 Nginx 依赖的 3 个包，若系统已安装，则可以忽略。代码如下：

```
yum install openssl
yum install zlib
yum install pcre
```

安装 Nginx：

```
nginx: yum install nginx
```

启动 Nginx：

```
systemctl start nginx.service
```

此时通过浏览器访问当前 IP 地址可以看到显示有“Welcome to nginx on Fedora!”的测试页，这个页面最后一句话“nginx configuration file /etc/nginx/nginx.conf.”提示了配置文件路径。

设置开机时自动启动：

```
systemctl enable nginx.service
```

6.7 SSL 免费证书申请

具体步骤如下。

（1）购买证书：在控制台的“产品与服务 → 安全（云盾）→SSL 证书”位置。单击“购买证书”按钮，选择 Symantec 的免费型 DV SSL。购买两个证书，购买后列表如图 6-3 所示。

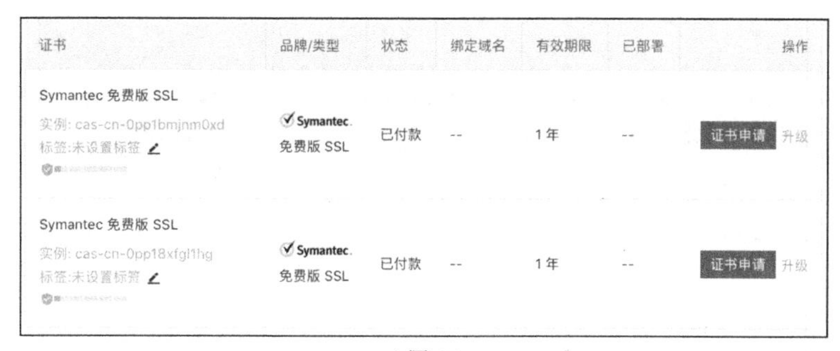

▲图 6-3

（2）申请证书：单击“证书申请”按钮，按要求填写域名等相关信息，如图 6-4 所示。

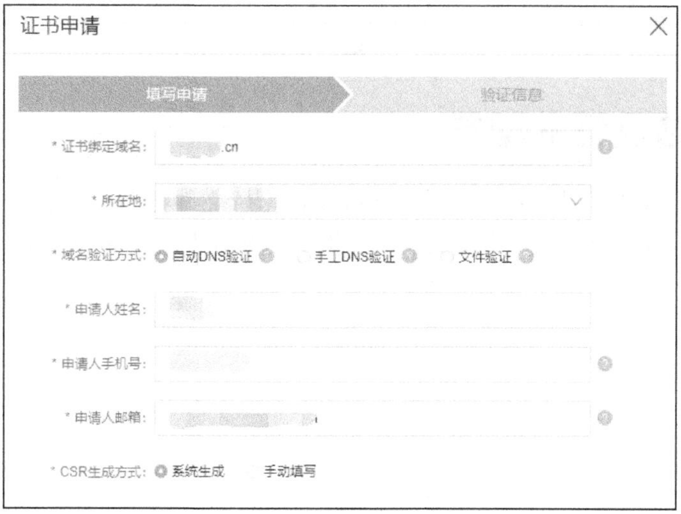

证书申请

填写申请　验证信息

* 证书绑定域名：.cn

* 所在地：

* 域名验证方式：自动DNS验证　手工DNS验证　文件验证

* 申请人姓名：

* 申请人手机号：

* 申请人邮箱：

* CSR生成方式：系统生成　手动填写

▲图 6-4

注意免费的 SSL 不能含通配符，例如“*”，即使是二级域名也要单独申请。例如 www.*XXX*.com 和 api.*XXX*.com 需要一共申请两个证书。注意：若域名只写了 *XXX*.com，则可用于 *XXX*.com 和 www. *XXX*.com 的访问，而不能用于 api. *XXX*.com。

（3）验证：填写完证书申请提交后，会进行域名验证。如果上一步“填写申请”中的“域名验证方式”选择的是“自动 DNS 验证”，则会自动在域名解析处添加一条验证信息。若选择了手工 NDS 验证，则需要自己按照提示添加验证信息。这条信息仅用于本次验证，验证签发之后，自己手动删除此条即可。验证信息如图 6-5 所示。

（4）提交审核：验证成功后，单击下面的按钮提交审核，然后等待审核通过。此步骤时最快需要 1 分钟，然后刷新页面，若发现进行到了“已签发”阶段，就可以下载使用证书了。

（5）下载证书：选择 Nginx 的证书，每个证书包含一个 key 文件和一个 pem 文件。在/etc/nginx 目录下新建一个 certs 文件夹（也可以放置在其他目录），将两个网站的证书（4 个文件）放进文件夹。

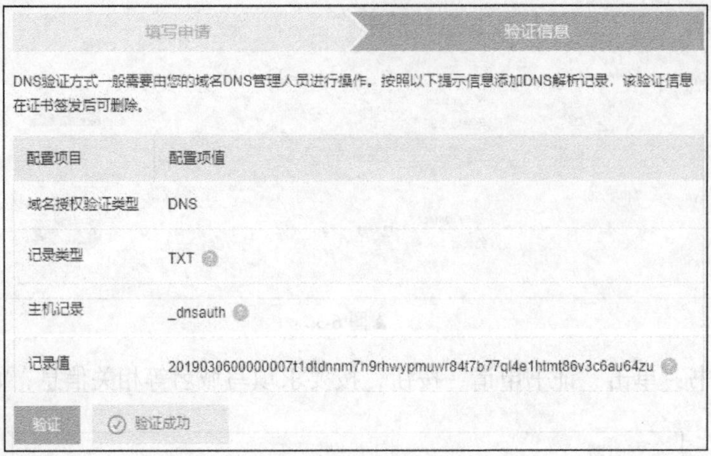

填写申请　验证信息

DNS验证方式一般需要由您的域名DNS管理人员进行操作。按照以下提示信息添加DNS解析记录，该验证信息在证书签发后可删除。

配置项目	配置项值
域名授权验证类型	DNS
记录类型	TXT
主机记录	_dnsauth
记录值	2019030600000007t1dtdnnm7n9rhwypmuwr84t7b77ql4e1htmt86v3c6au64zu

验证　验证成功

▲图 6-5

6.8 多网站及证书配置

在第二步安装 Nginx 时，我们知道了 Nginx 的配置文件在 /etc/nginx/nginx.conf 目录下，打开这个文件，可以看出它采用了类似 JSON 格式的配置方式，其配置结构类似如下样式：

```
http {

server {


    }


# Settings for a TLS enabled server.
#
#    server {
#        listen       443 ssl http2 default_server;
#        listen       [::]:443 ssl http2 default_server;
#        server_name  _;
#        root         /usr/share/nginx/html;
#    }

}
```

看到上面用“#”注释掉了 listen 443 ssl 的一个 server {}，这就是 SSL 的配置例子。我们可以在这里以这样的方式配置两个网站，但对于多网站配置的情况，建议将每一个网站的配置文件分别放置在一个文件中，方便后期修改维护。

在/etc/nginx 目录下新建一个 vhosts 文件夹，在其中新建名为 web.conf 和 api.conf 的两个文本文件，内容如下（以 web.conf 为例）：

```
    server {
        listen *:443   ssl;
        server_name  www.XXX.com;
        ssl_certificate /etc/ssl/certs/www/api/www.XXX.com.pem;
        ssl_certificate_key /etc/ssl/certs/www/api/www.XXX.com.key;
        ssl_protocols TLSv1.1 TLSv1.2;
        ssl_prefer_server_ciphers on;
        ssl_ciphers "EECDH+AESGCM:EDH+AESGCM:AES256+EECDH:AES256+EDH";
        ssl_ecdh_curve secp384r1;
        ssl_session_cache shared:SSL:10m;
        ssl_session_tickets off;
        ssl_stapling on; #ensure your cert is capable
        ssl_stapling_verify on; #ensure your cert is capable

        add_header Strict-Transport-Security "max-age=63072000; includeSubdomains;
preload";
        add_header X-Frame-Options DENY;
        add_header X-Content-Type-Options nosniff;

        #Redirects all traffic
        location / {
            proxy_pass  http://localhost:8080;
        }
    }
```

在这个文件中设置域名、证书位置，以及最后的网站地址，也就是将域名和之前配置好的两个网站分别绑定。若要两个 conf 文件均生效，需要将它们添加到 Nginx 的配置文件/etc/nginx/nginx.conf。编辑该文件，将其中的 server{}部分修改为如下代码：

```
    include /etc/nginx/vhosts/*.conf;
    server {
        listen       80;
        add_header Strict-Transport-Security max-age=15768000;
        return 301 https://$host$request_uri;
}
```

其中，include 语句用于将 vhosts 文件夹中的所有 conf 文件引用到当前文件中。修改 server{}中的内容，目的是使原本访问 80 端口的 HTTP 请求全部转为 HTTPS 请求，强制使用 HTTPS 协议。

重新启动 Nginx，分别访问两个网站的域名，均可以正常访问。

6.9 启用 ForwardedHeaders 中间件

虽然一切看起来都很好，但由于使用了 Nginx 作为反向代理，对于一些特殊需求还是会存在问题。例如需要获取访问者的 IP 地址，可以使用如下代码：

```
public IActionResult Index()
{
    ViewBag.IP = Request.HttpContext.Connection.RemoteIpAddress;
    return View();
}
```

但如果在 View 中将这个 IP 地址显示出来，可以看到这样的值：

```
::ffff:127.0.0.1
```

实际访问网站的是 Nginx，而网站和 Nginx 部署在同一服务器，这就获得了这样的访问者 IP 地址。请求 IP 地址一般放置在请求的 Header 中，如果想获取访问者的 IP 地址，需要 Nginx 把用户请求的 Header 中的一些其他信息也转发给网站，并做如下设置。

网站需要启用 ForwardedHeaders 中间件转发。在 Startup 的 Configure 中添加如下代码，注意 UseForwardedHeaders 要用在 UseAuthentication 之前。

```
app.UseForwardedHeaders(new ForwardedHeadersOptions
{
    ForwardedHeaders = ForwardedHeaders.XForwardedFor | ForwardedHeaders.XForwardedPr
oto
});

app.UseAuthentication();
```

修改 Nginx 的配置，新建一个名为 proxy.conf 的文本文件，内容为：

```
proxy_redirect          off;
proxy_set_header            Host          $host;
proxy_set_header            X-Real-IP          $remote_addr;
proxy_set_header            X-Forwarded-For $proxy_add_x_forwarded_for;
proxy_set_header                X-Forwarded-Proto $scheme;
client_max_body_size 10m;
client_body_buffer_size             128k;
proxy_connect_timeout     90;
proxy_send_timeout     90;
proxy_read_timeout     90;
proxy_buffers          32 4k;
```

将 proxy.conf 放置在 vhosts 文件夹中，因为已经设置了 nginx.conf 对这个文件夹的文件全部引用，因此也可以放置在其他位置，然后在 nginx.conf 中添加对这个文件的引用。

再次访问这个网站，可以看到访问者的真实 IP 地址。

6.10 独立部署（SCD）

下面来看独立部署（包含依赖项）的发布方式。上文的例子采用的是依赖框架部署，首先安装了.NET Core 和 ASP.NET Core 的 Runtime，这类似于之前部署.NET 相关应用时都需要安装.NET Framework。之所以在发布时产生的文件较少，是因为大部分依赖已经安装在目标系统中了；而独立部署，就是不需要安装 Runtime，直接将依赖一起发布出来，这会导致文件相当多。采用默认模板创建的 ASP.NET Core MVC 应用，不做修改直接发布会有 300 多个文件，大部分是 dll 文件。若采用独立部署发布，只需在发布时进行如图 6-6 所示的设置。

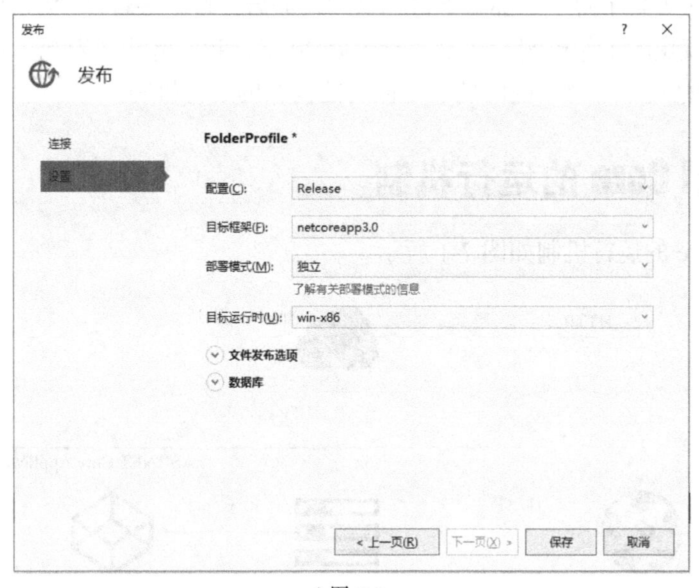

▲图 6-6

注意：目标运行时要与服务器的系统类型对应。

第7章　架构概览

“跨平台”后的 ASP.NET Core 是如何接收并处理请求的呢？它的运行和处理机制和之前有什么不同？就像大致了解了一部新手机的基本功能后，接下来将其“拆解”，看一下它的内部结构及各主要模块的工作机制。从第 7 章到第 20 章，我们将从“宏观”到“微观”来了解 ASP.NET Core 的结构，以及不同时期、不同功能模块的运行机制。

7.1 ASP.NET Core 的运行机制

ASP.NET Core 的运行机制如图 7-1 所示。

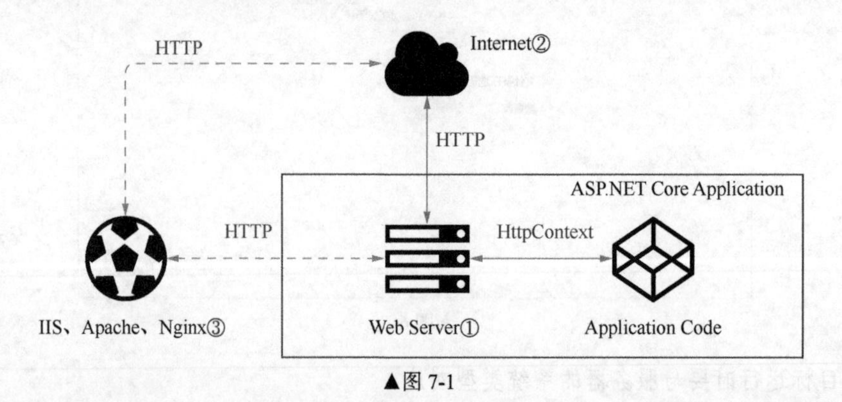

▲图 7-1

（1）Web Server：ASP.NET Core 提供两种服务器可用，分别是 Kestrel 和 HTTP.sys（Core 1.x 中被命名为 WebListener）。

- Kestrel 是一个跨平台的 Web 服务器。
- HTTP.sys 只能用在 Windows 操作系统中。

（2）Internet：当需要部署在 Internet 中并需要使用 Kestrel 中没有的功能（如 Windows 身份验证）时，可以选择 HTTP.sys。

（3）IIS、Apache、Nginx：Kestrel 可以单独使用，也可以将其与反向代理服务器（如 IIS、Nginx 或 Apache）结合使用。请求经这些服务器进行初步处理后转发给 Kestrel（即图中虚线的可选流程）。

大概的运行机制就是这样，那么具体到 ASP.NET Core Application 是如何运行的呢？我们将图 7-1 中 ASP.NET Core Application 所在的框中内容放大，看下一节。

7.2 ASP.NET Core Application 的架构

将图 7-1 的 ASP.NET Core Application 放大，架构如图 7-2 所示。

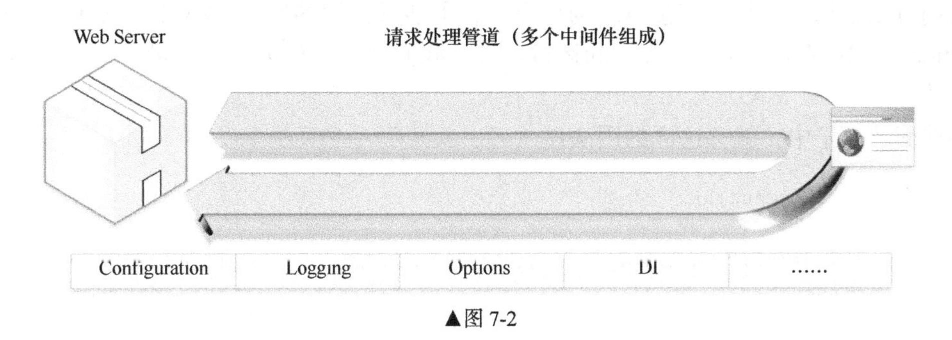

▲图 7-2

当请求到达 Web Server 后，被处理为 HttpContext，进一步交给请求处理管道来处理。这个管道由多个中间件组成，HttpContext 经管道处理后返回给用户。除了这个主要架构，还有一些辅助的功能模块，例如配置、日志、依赖注入系统等。

7.3 对 HTTP/2 的支持

从 ASP.NET Core 3.0 开始，在默认情况下，Kestrel 为 HTTPS 启用了 HTTP/2。当应用所在的操作系统支持 HTTP/2 时，将启用对 IIS 或 HTTP.sys 的 HTTP/2 支持。

以下运行机制中的 ASP.NET Core 支持 HTTP/2。

7.3.1 Kestrel

操作系统：Windows Server 2016/Windows 10 或更高版本；OpenSSL 1.0.2 或更高版本的 Linux（例如，Ubuntu 16.04 或更高版本）。

目标框架：.NET Core 2.2 或更高版本。

7.3.2 IIS（进程内）

操作系统：Windows Server 2016/Windows 10 或更高版本；IIS 10 或更高版本。

目标框架：.NET Core 2.2 或更高版本。

Kestrel 在 Windows Server 2012 R2 和 Windows 8.1 上对 HTTP/2 的支持受限。支持受限是因为在这些操作系统上可用的受支持 TLS 密码套件列表有限，可能需要使用椭圆曲线数字签

名算法 （ECDSA）生成的证书来保护 TLS 连接。

HTTP/2 连接必须使用应用层协议协商（ALPN）和 TLS 1.2 或更高版本。

7.4 ASP.NET Core 的环境变量

ASP.NET Core 中的环境变量非常重要且常用，它由 ASPNETCORE_ENVIRONMENT 变量指定。

我们可以根据需要将此变量设置为任意值，但通常使用的值是 Development、Staging 和 Production。它定义了当前应用程序的运行环境，我们经常会根据这个变量让应用采用不同的处理方式。

在前面的例子中，就有这样的用法：

```
if (env.IsDevelopment())
{
    var appAssembly = Assembly.Load(new AssemblyName(env.ApplicationName));
    if (appAssembly != null)
    {
        config.AddUserSecrets(appAssembly, optional: true);
    }
}
```

在 Layout View 中也有采用环境变量的代码：

```
<environment include="Development">
    <link rel="stylesheet" href="~/lib/bootstrap/dist/css/bootstrap.css" />
    <link rel="stylesheet" href="~/css/site.css" />
</environment>
```

因此，如果在应用程序运行之前将 ASPNETCORE_ENVIRONMENT 变量设置为 Development（或在 launchSettings.json 文件中设置此环境变量），应用程序会在 Development 模式下运行，而不是 Production 模式（这是不设置任何变量时的默认模式）下运行。

注意：在 Windows 和 macOS 操作系统上，环境变量和值不区分大小写，Linux 环境变量和值区分大小写。

第 8 章　应用启动

上一章从宏观的角度讲解了 ASP.NET Core 的整体架构，看完之后读者可能会有疑问，比如处理管道是怎么构建起来的？启动过程中，系统"默默地"做了哪些准备工作？本章我们详细介绍这个过程。

本章内容是按照从应用的启动运行到请求处理的顺序，比较符合正常的思维逻辑，但本章会涉及下文的一些知识点，如利用依赖注入、配置、Options 等。如果读者对这些知识不熟悉，可能会影响对本章内容的理解，建议先粗略了解本章的内容，待看完第 11～15 章后再精读本章。

8.1 概述

图 7-2 所示的大部分功能是在 ASP.NET Core Application 的启动阶段完成的，而这一切都在 Program 类中执行。因为 ASP.NET Core 应用程序本质上是控制台应用程序，所以它也是以一个我们熟悉的 Main 方法作为程序的起点。看一下这部分代码：

```
public class Program
    {
        public static void Main(string[] args)
        {
            CreateHostBuilder(args).Build().Run();
        }

        public static IHostBuilder CreateHostBuilder(string[] args) =>
            Host.CreateDefaultBuilder(args)
                .ConfigureWebHostDefaults(webBuilder =>
                {
                    webBuilder.UseStartup<Startup>();
                });
    }
```

Program 中定义了一个 CreateHostBuilder 方法，在 Main 方法中调用它返回一个 IHostBuilder，

并通过这个 IHostBuilder 的 Build 方法创建 IHost，接着调用 Run 方法使应用运行起来。图 8-1 简要描述了这一过程。

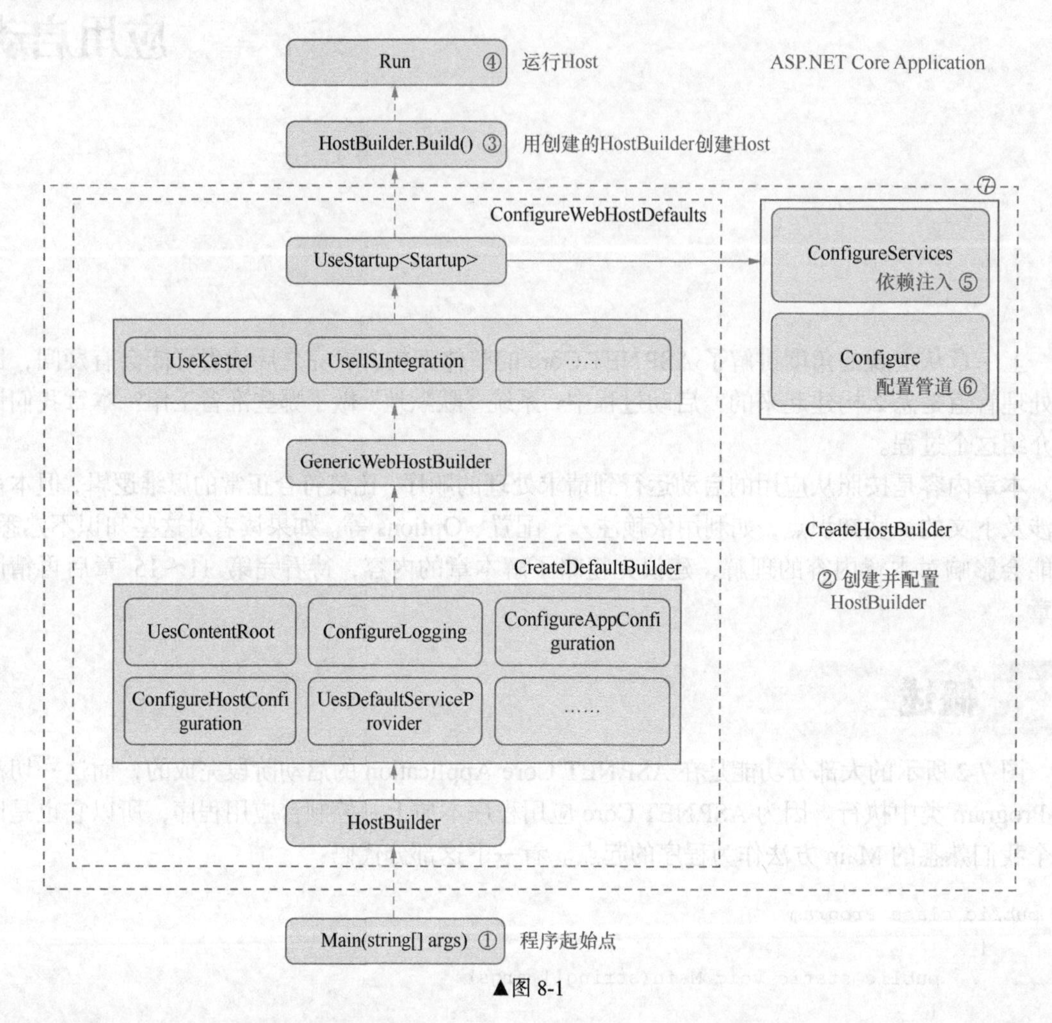

▲图 8-1

① Main 方法，是程序的起点。

② 创建并配置 IHostBuilder：首先调用 CreateHostBuilder 方法，包含 CreateDefaultBuilder 和 ConfigureWebHostDefaults 两个方法，如图 8-1 所示，它是一系列配置的大综合，下文将详细介绍。进行一系列配置后，调用 UseStartup<Startup>()，指定⑦Startup 文件为启动配置文件。在 Startup 文件中，将完成⑤服务的依赖注入和⑥配置管道的工作。

③ 生成 HostBuilder 并进行一系列配置后，通过 IHostBuilder 新建一个 Host 实例。

④ 调用 Host 的 Run 方法，并开始运行。

⑤ 依赖注入配置。

⑥ 请求处理管道的配置。

可以看出，应用的启动就是通过 Main 方法中的一行代码实现的：

```
CreateHostBuilder(args).Build().Run();
```

本章将对这行代码的 3 段内容逐一进行介绍。

8.2 HostBuilder 的创建与配置

对应 Main 方法的第一段：CreateHostBuilder(args)。其中涉及 Host.CreateDefaultBuilder 和 IHostBuilder.ConfigureWebHostDefaults 两个方法，看这两个方法的名字就知道，可以通过它们做一些默认的配置。

在 ASP.NET Core 1.0 中没有这样的方法，首次出现是在 ASP.NET Core 2.0 中，该版本新增了一个 CreateDefaultBuilder 方法，其目的是使代码更加简洁、使用更加方便，并且将一些在常规情况下需要调用的方法和配置放到了这个方法中。到了 3.0 版，又将 Web 相关的一些代码拆分出来放在了 IHostBuilder. ConfigureWebHostDefaults 中。

Host.CreateDefaultBuilder 方法创建一个 IHostBuilder 并对其进行配置，进而调用这个 IHostBuilder 的 ConfigureWebHostDefaults 方法进行一些 Web 相关的配置工作。

8.2.1 Host 的创建者 HostBuilder

Main 方法最先调用的 Host.CreateDefaultBuilder 的代码如下：

```
public static IHostBuilder CreateDefaultBuilder(string[] args)
{
var builder = new HostBuilder();

//省略了一些方法调用的代码

    return builder;
}
```

首先创建一个 HostBuilder，然后调用一些方法进行配置，最终返回这个 HostBuilder。可见，IHostBuilder 的具体实现就是 HostBuilder。HostBuilder 的代码如下：

```
public class HostBuilder : IHostBuilder
{
    private List<Action<IConfigurationBuilder>> _configureHostConfigActions = new List
<Action<IConfigurationBuilder>>();
    private List<Action<HostBuilderContext, IConfigurationBuilder>>
_configureAppConfigActions = new List<Action<HostBuilderContext, IConfigurationBuilder>>();
    private List<Action<HostBuilderContext, IServiceCollection>>
_configureServicesActions = new List<Action<HostBuilderContext, IServiceCollection>>();
    private List<IConfigureContainerAdapter> _configureContainerActions = new List
<IConfigureContainerAdapter>();
```

```
    private IServiceFactoryAdapter _serviceProviderFactory = new ServiceFactoryAdapter
<IServiceCollection>(new DefaultServiceProviderFactory());
    private bool _hostBuilt;
    private IConfiguration _hostConfiguration;
    private IConfiguration _appConfiguration;
    private HostBuilderContext _hostBuilderContext;
    private HostingEnvironment _hostingEnvironment;
    private IServiceProvider _appServices;

    public IDictionary<object, object> Properties { get; } = new Dictionary<object,
object>();

    public IHostBuilder ConfigureHostConfiguration(Action<IConfigurationBuilder>
configureDelegate)
    {
        _configureHostConfigActions.Add(configureDelegate ?? throw new ArgumentNullException
(nameof(configureDelegate)));
        return this;
    }

    public IHostBuilder ConfigureAppConfiguration(Action<HostBuilderContext,
IConfigurationBuilder> configureDelegate)
    {
        _configureAppConfigActions.Add(configureDelegate ?? throw new ArgumentNull
Exception(nameof(configureDelegate)));
        return this;
    }

    public IHostBuilder ConfigureServices(Action<HostBuilderContext, Iservice
Collection> configureDelegate)
    {
        _configureServicesActions.Add(configureDelegate ?? throw new
Argument NullException(nameof(configureDelegate)));
        return this;
    }

    public IHostBuilder UseServiceProviderFactory<TContainerBuilder>(IserviceProvider
Factory<TContainerBuilder> factory)
    {
        _serviceProviderFactory = new ServiceFactoryAdapter<TContainerBuilder>
(factory ?? throw new ArgumentNullException(nameof(factory)));
        return this;
    }

    public IHostBuilder UseServiceProviderFactory<TContainerBuilder>(Func<HostBuilder
Context, IServiceProviderFactory<TContainerBuilder>> factory)
```

```
    {
        _serviceProviderFactory = new ServiceFactoryAdapter<TContainerBuilder>(() =>
_hostBuilderContext, factory ?? throw new ArgumentNullException(nameof(factory)));
        return this;
    }

    public IHostBuilder ConfigureContainer<TContainerBuilder>(Action<HostBuilderContext,
 TContainerBuilder> configureDelegate)
    {
        _configureContainerActions.Add(new ConfigureContainerAdapter<TContainerBuilder>
(configureDelegate
            ?? throw new ArgumentNullException(nameof(configureDelegate))));
        return this;
}

//暂时省略后面的代码，下文会讲到
}
```

这只是本节会用到的前半部分代码，包含几个类型为 Action 的集合和对应的处理方法。图 8-2 显示了它的组成成员。

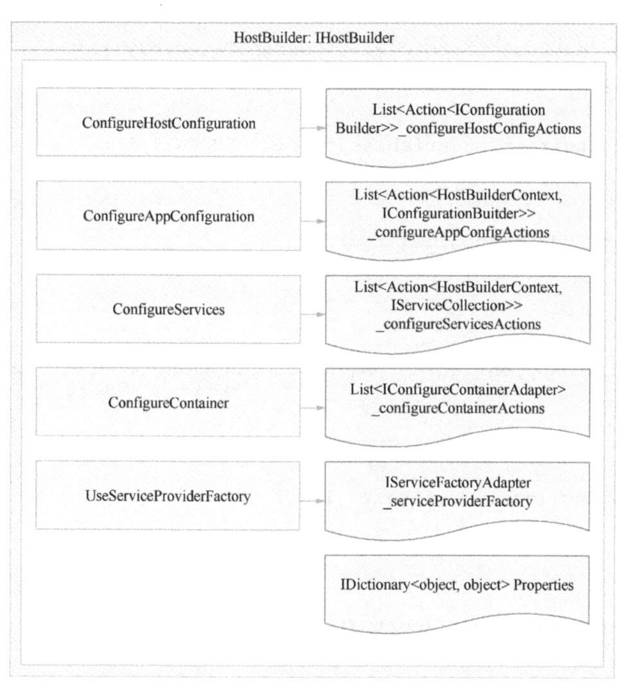

▲图 8-2

这些成员的功能如下。

- **_configureHostConfigActions**：与 HostBuilder 自身配置相关的一些 Action 的集合。

- **_configureAppConfigActions**：对应用进行配置的一些 Action 的集合。
- **_configureServicesActions**：用于将服务注册到依赖注入容器中的一些 Action 的集合，这在第 10 章会讲到。
- **_configureContainerActions 和 _serviceProviderFactory**：依赖注入容器构建过程中用到的 Actions 和 Factory（ASP.NET Core 3.0 开始，更改了替换默认容器的方案）。
- **Properties**：用于 HostBuilder 配置和创建 Host 各种组件之间的数据共享的字典。

而 HostBuilder 的 ConfigureHostConfiguration、ConfigureAppConfiguration、ConfigureServices 等方法恰好对应的是它的这几个成员，通过这几个方法向对应的成员插入对应的 Action。图 8-2 显示了这些成员和方法的一对一（每行）关系。

注意：在 Microsoft.Extensions.Hosting 和 Microsoft.Extensions.Hosting.Internal 中各存在一个名为 Host 的类，此处用到的是前者，内容是一些 HostBuilder 的配置方法；后者是 ASP.NET 的一个核心类，由 HostBuilder 创建。

Host.CreateDefaultBuilder 方法的代码如下：

```
public static IHostBuilder CreateDefaultBuilder(string[] args)
{
    var builder = new HostBuilder();

    builder.UseContentRoot(Directory.GetCurrentDirectory());
    builder.ConfigureHostConfiguration(config =>
    {
        config.AddEnvironmentVariables(prefix: "DOTNET_");
        if (args != null)
        {
            config.AddCommandLine(args);
        }
    });

    builder.ConfigureAppConfiguration((hostingContext, config) =>
    {
        //省略代码
    })
    .ConfigureLogging((hostingContext, logging) =>
    {
        //省略代码
    })
    .UseDefaultServiceProvider((context, options) =>
    {
        //省略代码
    });

    return builder;
}
```

中间省略的代码是一些 builder.Configure*XXX* 和 builder.Use*XXX* 方法，其并不是继承自 HostBuilder，而是在 HostingHostBuilderExtensions 中做的扩展。例如第一个 UseContentRoot 方法代码如下：

```
public static IHostBuilder UseContentRoot(this IHostBuilder hostBuilder, string contentRoot)
{
    return hostBuilder.ConfigureHostConfiguration(configBuilder =>
    {
        configBuilder.AddInMemoryCollection(new[]
        {
            new KeyValuePair<string, string>(HostDefaults.ContentRootKey,
contentRoot  ?? throw new ArgumentNullException(nameof(contentRoot)))
        });

    });
}
```

这个 UseContentRoot 方法正是调用了 HostBuilder 的 ConfigureHostConfiguration 方法，也就是将 UseContentRoot 方法内部定义的 Action 写入 HostBuilder 的_configureHostConfigActions 集合中。

其他几个方法同理，这里就不逐一说明了，图 8-3 描述了上述过程。

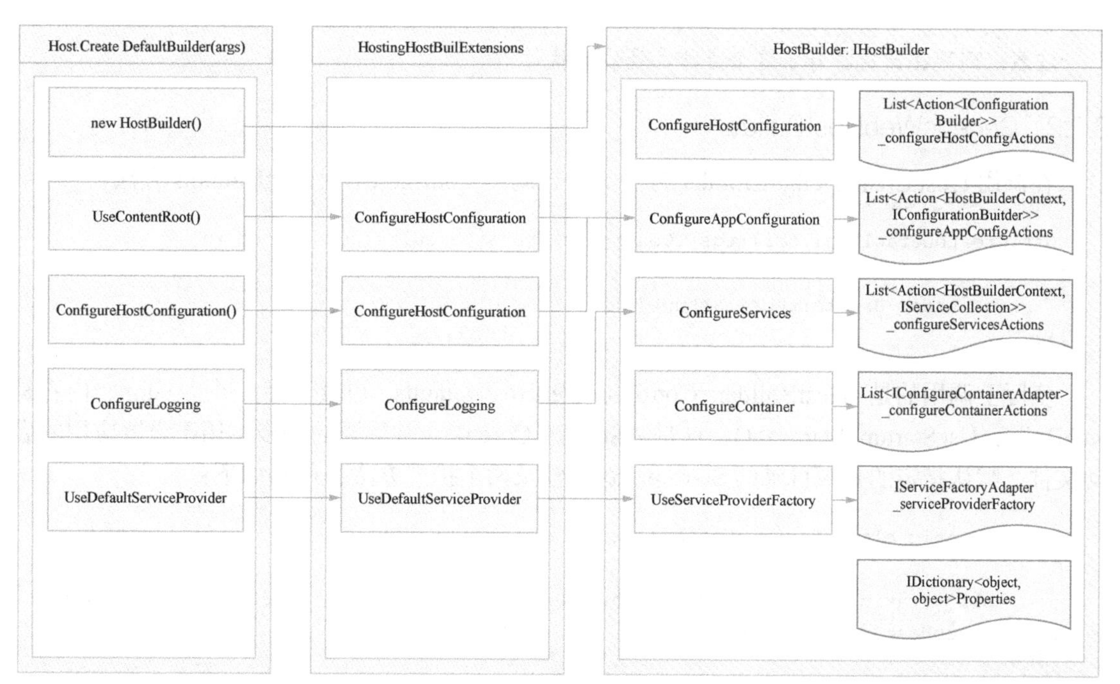

▲图 8-3

下面介绍这些方法的作用。

- **UseContentRoot**：为应用程序指定根目录。需要注意这和 StaticFiles 的根目录是不同的，虽然默认情况下 StaticFiles 的根目录以 ContentRoot 为依据（[ContentRoot]/wwwroot）。
- **ConfigureHostConfiguration 和 ConfigureAppConfiguration**：读取配置信息。前者针对 HostBuilder 的配置，后者针对应用的配置。默认情况下会读取 appsettings.json 和 appsettings.{env.EnvironmentName}.json 中的配置，env.EnvironmentName 指环境变量，例如 Development 等，如下代码所示。当在 Development 环境下运行时，还会读取用户密钥，这部分将在学习系统配置时详细介绍。
- **ConfigureLogging**：配置日志处理程序、控制台和调试日志提供程序，这部分在学习日志时详细讲解。
- **UseDefaultServiceProvider**：设置默认的依赖注入容器，这部分在学习依赖注入时详细讲解。

这些方法都将自身涉及的配置操作作为 Action 写入 HostBuilder 的对应集合中，就像我们写代码一样，写了如下几种方法：

```
Build(){
HostBuilder. ConfigureHostFunction1().ConfigureHostFunction2()…;
HostBuilder. ConfigureAppFunction1().ConfigureAppFunction2()…;
……
}
```

注意：到现在只是写好了这个方法，还没有执行。

8.2.2 GenericWebHostBuilder

在调用 Host.CreateDefaultBuilder 方法之后，调用 ConfigureWebHostDefaults 方法：

```
Host.CreateDefaultBuilder(args).ConfigureWebHostDefaults(webBuilder =>
{
    webBuilder.UseStartup<Startup>();
});
```

实际上调用的是 IHostBuilder. ConfigureWebHostDefaults，并传入了一个 Action，内容是 webBuilder.UseStartup<Startup>()。这是 ASP.NET Core 中非常重要的一步，用于设置应用的启动文件，默认指定的是根目录的 Startup。这里涉及两个扩展方法，代码如下：

```
public static class GenericHostBuilderExtensions
{
    public static IHostBuilder ConfigureWebHostDefaults(this IHostBuilder builder,
Action<IWebHostBuilder> configure)
    {
        return builder.ConfigureWebHost(webHostBuilder =>
        {
```

```
                WebHost.ConfigureWebDefaults(webHostBuilder);

                configure(webHostBuilder);
            });
        }
    }
    public static class GenericHostWebHostBuilderExtensions
    {
        public static IHostBuilder ConfigureWebHost(this IHostBuilder builder, Action<IWe
bHostBuilder> configure)
        {
            var webhostBuilder = new GenericWebHostBuilder(builder);
            configure(webhostBuilder);
            builder.ConfigureServices((context, services) => services.AddHostedService
<GenericWebHostService>());
            return builder;
        }
    }
```

这两个方法的功能如下。

（1）创建一个 GenericWebHostBuilder。从其名字看出主要涉及 Web 相关的内容，简化后的代码如下：

```
internal class GenericWebHostBuilder : IWebHostBuilder, ISupportsStartup,
Isupports UseDefaultServiceProvider
{
    private readonly IHostBuilder _builder;
    private readonly IConfiguration _config;
    private readonly object _startupKey = new object();

    private AggregateException _hostingStartupErrors;
    private HostingStartupWebHostBuilder _hostingStartupWebHostBuilder;


    public GenericWebHostBuilder(IHostBuilder builder)
    {
        _builder = builder;

        _config = new ConfigurationBuilder()
            .AddEnvironmentVariables(prefix: "ASPNETCORE_")
            .Build();

        _builder.ConfigureHostConfiguration(config =>
        {
            config.AddConfiguration(_config);
            ExecuteHostingStartups();
```

```
        });
        //省略几个 Configure 方法
    }

    public IWebHostBuilder ConfigureAppConfiguration(Action<WebHostBuilderContext,
IConfigurationBuilder> configureDelegate)
    {
        _builder.ConfigureAppConfiguration((context, builder) =>
        {
            var webhostBuilderContext = GetWebHostBuilderContext(context);
            configureDelegate(webhostBuilderContext, builder);
        });

        return this;
    }

//省略几个 Configure 方法

    public IWebHostBuilder ConfigureServices(Action<WebHostBuilderContext,
Iservice Collection> configureServices)
    {
        _builder.ConfigureServices((context, builder) =>
        {
            var webhostBuilderContext = GetWebHostBuilderContext(context);
            configureServices(webhostBuilderContext, builder);
        });

        return this;
    }

    public IWebHostBuilder UseStartup(Type startupType)
    {
        // 省略代码，后文介绍
    }

    private void UseStartup(Type startupType, HostBuilderContext context,
Iservice Collection services)
    {
        // 省略代码，后文介绍
}
}
```

除了最后的两个 UseStartup 方法，从它的构造方法和提供的 ConfigureAppConfiguration、ConfigureServices 方法来看，都是在调用它的 IHostBuilder _builder 成员的对应方法，而这个成员就是上文的 HostBuilder，所以 GenericWebHostBuilder 有点像是对 HostBuilder 的封装，实际

作用同样是向 HostBuilder 的几个集合中添加 Action，只不过侧重点都是与 Web 相关的内容。最后的两个 UseStartup 方法将在稍后详细介绍。

（2）创建 GenericWebHostBuilder 之后，会调用 WebHost.ConfigureWebDefaults(webHostBuilder)，并对其进行配置：

```
internal static void ConfigureWebDefaults(IWebHostBuilder builder)
{
    builder.ConfigureAppConfiguration((ctx, cb) =>
    {
        if (ctx.HostingEnvironment.IsDevelopment())
        {
            StaticWebAssetsLoader.UseStaticWebAssets(ctx.HostingEnvironment,
ctx.Configuration);
        }
    });
    builder.UseKestrel((builderContext, options) =>
    {
        options.Configure(builderContext.Configuration.GetSection("Kestrel"));
    })
    .ConfigureServices((hostingContext, services) =>
    {
        //省略部分代码

    })
    .UseIIS();
    .UseIISIntegration();
}
```

同样是一些 Configure*XXX* 和 builder.Use*XXX* 方法，它们的功能如下。

- **UseKestrel**：指定服务器使用 Kestrel，若使用 HTTP.sys，需使用 UseHttpSys。
- **UseIIS 和 UseIISIntegration**：将应用程序配置为在 IIS 中运行。这里仍需要使用 UseKestrel，IIS 起到反向代理的作用，Kestrel 仍用作主机。如果应用程序没有使用 IIS 作为反向代理，那么 UseIISIntegration 不会有任何效果。因此，即使应用程序在非 IIS 方案中运行，也可以安全调用这种方法。

（3）调用 GenericWebHostBuilder.UseStartup<Startup>()，指定启动文件，注意这里只是指定，而不是执行。这部分内容很重要，将在下一小节单独介绍。

（4）添加对 GenericWebHostService 的依赖注入方法。

总结：本小节主要涉及 ConfigureWebHostDefaults 方法中的内容，其功能主要是创建一个专注于 Web 服务的 GenericWebHostBuilder，并利用它对 HostBuilder 进行进一步的配置，逻辑类似图 8-3，同样是向 HostBuilder 添加一系列的 Action。

8.2.3　处理 Startup 文件

介绍 GenericWebHostBuilder 时略过了两个 UseStartup 方法，它们的作用主要是处理

Startup。Startup 是一个非常重要的启动配置文件，代码默认如下：

```
public class Startup
{
    public Startup(IConfiguration configuration)
    {
        Configuration = configuration;
    }

    public IConfiguration Configuration { get; }
    public void ConfigureServices(IServiceCollection services)
    {
        services.AddControllersWithViews();
    }
    public void Configure(IApplicationBuilder app, IWebHostEnvironment env)
    {
            //用于构建请求处理管道，代码省略
    }
}
```

Startup 默认有两个方法——ConfigureServices 和 Configure。

注意：查看代码发现多了一个 ConfigureContainer 方法，是可选项。如果想采用其他的依赖注入的容器替换默认容器，可以新增这个方法用于依赖注入的注册。

对于一些需要注册到依赖注入容器中的内容，可以在 ConfigureServices 方法中进行配置，例如默认代码中的 services.AddControllersWithViews()方法可以将一些 Controller 和 View 相关的组件注册到容器中。Configure 方法用于构建请求处理管道，会在下文进行详细介绍。

GenericWebHostBuilder 的两个 UseStartup 方法的代码如下：

```
internal class GenericWebHostBuilder : IWebHostBuilder, ISupportsStartup,
Isupports UseDefaultServiceProvider
{
    private readonly IHostBuilder _builder;
    private readonly IConfiguration _config;
    private readonly object _startupKey = new object();

    public IWebHostBuilder UseStartup(Type startupType)
    {
        //
        _builder.Properties["UseStartup.StartupType"] = startupType;
        _builder.ConfigureServices((context, services) =>
        {
            if (_builder.Properties.TryGetValue("UseStartup.StartupType", out var
cachedType) && (Type)cachedType == startupType)
            {
                UseStartup(startupType, context, services);
```

```
            }
        });

        return this;
}
}

private void UseStartup(Type startupType, HostBuilderContext context, Iservice
Collection services)
{
    var webHostBuilderContext = GetWebHostBuilderContext(context);
    var webHostOptions = (WebHostOptions)context.Properties[typeof(WebHostOptions)];

    ExceptionDispatchInfo startupError = null;
    object instance = null;
    ConfigureBuilder configureBuilder = null;

    try
    {
        if (typeof(IStartup).IsAssignableFrom(startupType))
        {
            throw new NotSupportedException($"{typeof(IStartup)} isn't supported");
        }
        if (StartupLoader.HasConfigureServicesIServiceProviderDelegate(startupType,
context.HostingEnvironment.EnvironmentName))
        {
            throw new NotSupportedException($"ConfigureServices returning an
{typeof(IServiceProvider)} isn't supported.");
        }

        instance = ActivatorUtilities.CreateInstance(new HostServiceProvider(webHost
BuilderContext), startupType);
        context.Properties[_startupKey] = instance;

        //查找 Startup.ConfigureServices 方法并执行
        var configureServicesBuilder = StartupLoader.FindConfigureServicesDelegate
(startupType, context.HostingEnvironment.EnvironmentName);
        var configureServices = configureServicesBuilder.Build(instance);

        configureServices(services);

        // Startup.ConfigureContainer 方法处理代码略

        //查找 Startup.Configure 方法
        configureBuilder = StartupLoader.FindConfigureDelegate(startupType,
context.HostingEnvironment.EnvironmentName);
```

```
        }
        catch (Exception ex) when (webHostOptions.CaptureStartupErrors)
        {
            startupError = ExceptionDispatchInfo.Capture(ex);
        }

        // 将 Startup.Configure 方法写入 Options，现在不执行
        services.Configure<GenericWebHostServiceOptions>(options =>
        {
            options.ConfigureApplication = app =>
            {
                startupError?.Throw();

                // 执行 Startup.Configure
                if (instance != null && configureBuilder != null)
                {
                    configureBuilder.Build(instance)(app);
                }
            };
        });
    }
```

可以看出第一个 UseStartup 方法此时会被执行，作用是将第二个 UseStartup 方法作为 Action 的一部分写入 HostBuilder 的_configureServicesActions 中。所以第二个 UseStartup 此时并未被执行，而会在下一节 Host 的构建时被执行。

为了便于理解，提前来看第二个方法的作用，主要是根据 UseStartup<Startup>()找到对应的 Startup，并判断 Startup 的有效性，进而识别出它的 configureServices、configureContainer 和 Configure 方法，代码中给出了相应的注释。

这里依赖注入了 StartupLoader 的 HasConfigureServicesIServiceProviderDelegate、FindConfigureDelegate、FindConfigureServicesDelegate 等方法，代码如下：

```
internal class StartupLoader
{
    //省略部分代码

    internal static ConfigureServicesBuilder FindConfigureServicesDelegate(Type
startupType, string environmentName)
    {
        var servicesMethod = FindMethod(startupType, "Configure{0}Services",
environmentName, typeof(IServiceProvider), required: false)
            ?? FindMethod(startupType, "Configure{0}Services", environmentName,
typeof(void), required: false);
        return new ConfigureServicesBuilder(servicesMethod);
    }
```

```
    private static MethodInfo FindMethod(Type startupType, string methodName, string
environmentName, Type returnType = null, bool required = true)
    {
        var methodNameWithEnv = string.Format(CultureInfo.InvariantCulture, methodName,
 environmentName);
        var methodNameWithNoEnv = string.Format(CultureInfo.InvariantCulture,
methodName, "");

        var methods = startupType.GetMethods(BindingFlags.Public | BindingFlags.
Instance | BindingFlags.Static);
        var selectedMethods = methods.Where(method =>
method.Name.Equals(methodName WithEnv, StringComparison.OrdinalIgnoreCase)).ToList();
        //验证代码省略
    }
}
```

这里只保留查找 ConfigureServices 方法的例子，其他方法也是类似。可以通过字符串查找“ConfigureServices”“Configure{ environmentName}Services”，最终查找到后会被执行。

找到 Startup.Configure 方法并不会立即执行，而是存储为 options.ConfigureApplication，等待执行时机。

总结：至此，通过 Host.CreateDefaultBuilder 和 IHostBuilder.ConfigureWebHostDefaults 两个默认配置方法将 HostBuilder 创建并配置完成，并在 HostBuilder 的几个集合类型的成员中写入了大量的 Action。一般情况下，调用这两个方法执行默认配置就足够用了，但既然这是默认配置，我们就可以根据自身情况自定义。因为这些配置都是对 HostBuilder 进行修改，然后返回修改后的 HostBuilder，所以在这两个方法无法满足实际需求的情况下，可以通过如下方法进行自定义。例如，通过将监听端口改为 8080 实现：

```
        public static IHostBuilder CreateHostBuilder(string[] args) =>
            Host.CreateDefaultBuilder(args)
                .ConfigureWebHostDefaults(webBuilder =>
                {
                    webBuilder.UseUrls("http://*:8080").UseStartup<Startup>();
                });
```

8.3 Host 的构建

对应 Main 方法的第二段：Build()。上一节创建并配置好了 HostBuilder，接下来通过它的 Build 方法创建 Host。上一节省略了 HostBuilder 的部分代码，而现在会用到，代码如下：

```
public class HostBuilder : IHostBuilder
{
    private List<Action<IConfigurationBuilder>> _configureHostConfigActions = new
List<Action<IConfigurationBuilder>>();
```

```
    private List<Action<HostBuilderContext, IConfigurationBuilder>> _configureApp
ConfigActions = new List<Action<HostBuilderContext, IConfigurationBuilder>>();
    private List<Action<HostBuilderContext, IServiceCollection>> _configureServices
Actions = new List<Action<HostBuilderContext, IServiceCollection>>();
    private List<IConfigureContainerAdapter> _configureContainerActions = new List
<IConfigureContainerAdapter>();
    private IServiceFactoryAdapter _serviceProviderFactory = new ServiceFactoryAdapter
<IServiceCollection>(new DefaultServiceProviderFactory());
    private bool _hostBuilt;
    private IConfiguration _hostConfiguration;
    private IConfiguration _appConfiguration;
    private HostBuilderContext _hostBuilderContext;
    private HostingEnvironment _hostingEnvironment;
    private IServiceProvider _appServices;

    public IDictionary<object, object> Properties { get; } = new Dictionary<object,
object>();

    public IHost Build()
    {
        if (_hostBuilt)
        {
            throw new InvalidOperationException("Build can only be called once.");
        }
        _hostBuilt = true;

        BuildHostConfiguration();
        CreateHostingEnvironment();
        CreateHostBuilderContext();
        BuildAppConfiguration();
        CreateServiceProvider();

        return _appServices.GetRequiredService<IHost>();
    }

    private void BuildHostConfiguration()
    {
        var configBuilder = new ConfigurationBuilder()
            .AddInMemoryCollection();

        foreach (var buildAction in _configureHostConfigActions)
        {
            buildAction(configBuilder);
        }
        _hostConfiguration = configBuilder.Build();
    }
```

```
    private void CreateHostingEnvironment()
    {
        _hostingEnvironment = new HostingEnvironment()
        {
            ApplicationName = _hostConfiguration[HostDefaults.ApplicationKey],
            EnvironmentName = _hostConfiguration[HostDefaults.EnvironmentKey] ??
Environments.Production,
            ContentRootPath = ResolveContentRootPath(_hostConfiguration[HostDefaults.
ContentRootKey], AppContext.BaseDirectory),
        };

        if (string.IsNullOrEmpty(_hostingEnvironment.ApplicationName))
        {
            _hostingEnvironment.ApplicationName = Assembly.GetEntryAssembly()?.GetName().
Name;
        }

        _hostingEnvironment.ContentRootFileProvider = new PhysicalFileProvider
(_hostingEnvironment.ContentRootPath);
    }

    private string ResolveContentRootPath(string contentRootPath, string basePath)
    {
        if (string.IsNullOrEmpty(contentRootPath))
        {
            return basePath;
        }
        if (Path.IsPathRooted(contentRootPath))
        {
            return contentRootPath;
        }
        return Path.Combine(Path.GetFullPath(basePath), contentRootPath);
    }

    private void CreateHostBuilderContext()
    {
        _hostBuilderContext = new HostBuilderContext(Properties)
        {
            HostingEnvironment = _hostingEnvironment,
            Configuration = _hostConfiguration
        };
    }

    private void BuildAppConfiguration()
    {
```

```
        var configBuilder = new ConfigurationBuilder()
            .SetBasePath(_hostingEnvironment.ContentRootPath)
            .AddConfiguration(_hostConfiguration, shouldDisposeConfiguration: true);

        foreach (var buildAction in _configureAppConfigActions)
        {
            buildAction(_hostBuilderContext, configBuilder);
        }
        _appConfiguration = configBuilder.Build();
        _hostBuilderContext.Configuration = _appConfiguration;
    }

    private void CreateServiceProvider()
    {
        var services = new ServiceCollection();
#pragma warning disable CS0618 //
        services.AddSingleton<IHostingEnvironment>(_hostingEnvironment);
#pragma warning restore CS0618 //
        services.AddSingleton<IHostEnvironment>(_hostingEnvironment);
        services.AddSingleton(_hostBuilderContext);
        // 以工厂的方式注册应用的配置
        services.AddSingleton(_ => _appConfiguration);
#pragma warning disable CS0618 //
        services.AddSingleton<IApplicationLifetime>(s =>
(IApplicationLifetime)s.GetService<IHostApplicationLifetime>());
#pragma warning restore CS0618 //
        services.AddSingleton<IHostApplicationLifetime, ApplicationLifetime>();
        services.AddSingleton<IHostLifetime, ConsoleLifetime>();
        services.AddSingleton<IHost, Internal.Host>();
        services.AddOptions();
        services.AddLogging();

        foreach (var configureServicesAction in _configureServicesActions)
        {
            configureServicesAction(_hostBuilderContext, services);
        }

        var containerBuilder = _serviceProviderFactory.CreateBuilder(services);

        foreach (var containerAction in _configureContainerActions)
        {
            containerAction.ConfigureContainer(_hostBuilderContext, containerBuilder);
        }

        _appServices = _serviceProviderFactory.CreateServiceProvider(containerBuilder
);
```

```
        if (_appServices == null)
        {
            throw new InvalidOperationException($"The IServiceProviderFactory returned
 a null IServiceProvider.");
        }

        _ = _appServices.GetService<IConfiguration>();
    }
}
```

代码有些长，图 8-4 的右半部分描述了这一过程：HostBuilder 的 Build 方法调用了它的 BuildHostConfiguration、CreateHostingEnvironment、CreateHostBuilderContext、BuildAppConfiguration、CreateServiceProvider 方法，这些方法实际上就是调用 HostBuilder 的创建与配置阶段配置的 Action 集合，例如 BuildHostConfiguration 方法通过遍历_configureHostConfigActions 集合来处理 ConfigurationBuilder，最终由 ConfigurationBuilder 生成 HostBuilder 的成员_hostConfiguration。

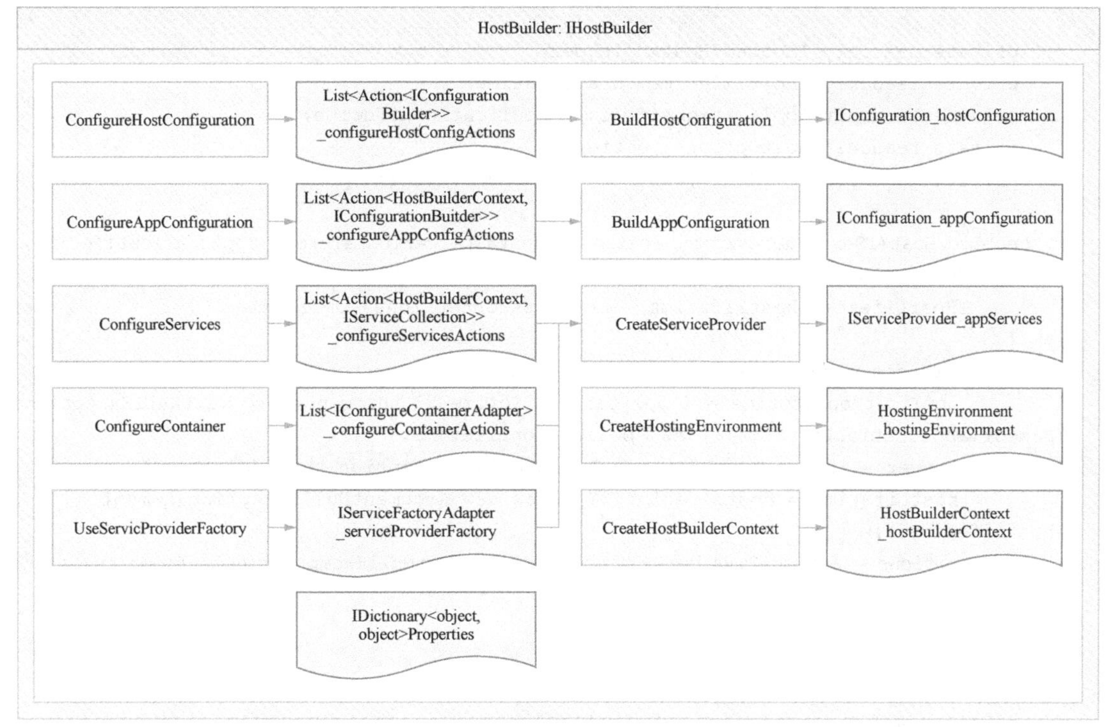

▲图 8-4

在最后执行的 CreateServiceProvider 方法中，生成的 HostBuilder 的_hostingEnvironment、_appConfiguration 等成员被注册到了依赖注入容器，同时注册的还有 ApplicationLifetime、Options 和 Logging 等。

在此也执行了 HostBuilder 的_configureServicesActions 集合中的 Action，其中包括 Startup 的处理，这在上一节已经讲过，此时会执行第二个 UseStartup 方法，获取 Startup 的实例并执行这个实例的 ConfigureServices 方法。这就是 ASP.NET Core 留给我们向依赖注入容器中注册组件的方法。

值得关注的是，此时也向依赖注入容器中注册了 IHost 类型，对应实现的是 Microsoft.Extensions.Hosting.Internal.Host：

```
services.AddSingleton<IHost, Internal.Host>();
```

在 HostBuilder 的 Build 方法最后，又从容器中获取了该 Host：

```
return _appServices.GetRequiredService<IHost>();
```

容器中获取该 Host 的目的是由容器创建 Host，且在创建时自然会提供 Host 中需要从依赖注入中获取的成员。Host 的代码如下：

```
internal class Host : IHost
{
    private readonly ILogger<Host> _logger;
    private readonly IHostLifetime _hostLifetime;
    private readonly ApplicationLifetime _applicationLifetime;
    private readonly HostOptions _options;
    private IEnumerable<IHostedService> _hostedServices;

    public Host(IServiceProvider services, IHostApplicationLifetime applicationLifetime,
 ILogger<Host> logger,
        IHostLifetime hostLifetime, IOptions<HostOptions> options)
    {
        Services = services ?? throw new ArgumentNullException(nameof(services));
        _applicationLifetime = (applicationLifetime ?? throw new ArgumentNullException
(nameof(applicationLifetime))) as ApplicationLifetime;
        _logger = logger ?? throw new ArgumentNullException(nameof(logger));
        _hostLifetime = hostLifetime ?? throw new ArgumentNullException(nameof
(hostLifetime));
        _options = options?.Value ?? throw new ArgumentNullException(nameof(options));
    }

    public IServiceProvider Services { get; }

    public async Task StartAsync(CancellationToken cancellationToken = default)
    {
        //方法内容省略，见下一节
    }
    //省略部分代码
}
```

其构造方法中注入了这些组件并赋值给了自身的成员变量，所以 HostBuilder 的 Build 方法最终返回一个 Internal.Host 实例，这个实例包含了由 HostBuilder 创建的多种组件。

至此，Host 构建完毕。从代码中可以看到，Host 还有一个 StartAsync 方法，下一节从这里开始讲 Host 的启动。

8.4 Host 的启动

对应 Main 方法的第三段：Run()。Host 构建完毕，下一步就是将它“Run”（运行）起来。Run 方法写在 HostingAbstractionsHostExtensions 中，代码如下：

```
public static class HostingAbstractionsHostExtensions
{
    //省略部分代码
    public static void Run(this IHost host)
    {
        host.RunAsync().GetAwaiter().GetResult();
    }

    public static async Task RunAsync(this IHost host, CancellationToken token = default)
    {
        try
        {
            await host.StartAsync(token);

            await host.WaitForShutdownAsync(token);
        }
        finally
        {
            {
                host.Dispose();
            }

        }
    }
}
```

它再次调用了 Host 的 StartAsync 方法，上一节最后的 Host 的代码中省略了这个方法，代码如下：

```
    public async Task StartAsync(CancellationToken cancellationToken = default)
    {
        _logger.Starting();

        using var combinedCancellationTokenSource = CancellationTokenSource.
CreateLinkedTokenSource(cancellationToken, _applicationLifetime.ApplicationStopping);
```

```
        var combinedCancellationToken = combinedCancellationTokenSource.Token;

        await _hostLifetime.WaitForStartAsync(combinedCancellationToken);

        combinedCancellationToken.ThrowIfCancellationRequested();
        _hostedServices = Services.GetService<IEnumerable<IHostedService>>();

        foreach (var hostedService in _hostedServices)
        {
            await
hostedService.StartAsync(combinedCancellationToken).ConfigureAwait(false);
        }

        _applicationLifetime?.NotifyStarted();

        _logger.Started();
    }
```

核心代码是从依赖注入容器中获取注册的IHostedService集合，然后遍历它们并逐一调用它们的StartAsync方法运行。

默认情况下会获取两个IhostedService：一个是DataProtectionHostedService，用于数据保护的服务；另一个是在处理GenericWebHostBuilder时注入容器中的GenericWebHostService。代码如下（省略了部分方法和用于验证的代码）：

```
internal class GenericWebHostService : IHostedService
{
    //省略构造方法代码
    public GenericWebHostServiceOptions Options { get; }
    public IServer Server { get; }
    public ILogger Logger { get; }
    public ILogger LifetimeLogger { get; }
    public DiagnosticListener DiagnosticListener { get; }
    public IHttpContextFactory HttpContextFactory { get; }
    public IApplicationBuilderFactory ApplicationBuilderFactory { get; }
    public IEnumerable<IStartupFilter> StartupFilters { get; }
    public IConfiguration Configuration { get; }
    public IWebHostEnvironment HostingEnvironment { get; }

    public async Task StartAsync(CancellationToken cancellationToken)
    {
        HostingEventSource.Log.HostStart();
        var serverAddressesFeature = Server.Features?.Get<IServerAddressesFeature>();
        var addresses = serverAddressesFeature?.Addresses;

        RequestDelegate application = null;
```

```
        try
        {
            Action<IApplicationBuilder> configure = Options.ConfigureApplication;

            var builder = ApplicationBuilderFactory.CreateBuilder(Server.Features);

            foreach (var filter in StartupFilters.Reverse())
            {
                configure = filter.Configure(configure);
            }

            configure(builder);

            application = builder.Build();
        }
        catch (Exception ex)
        {

        }

        var httpApplication = new HostingApplication(application, Logger, Diagnostic
Listener, HttpContextFactory);

        await Server.StartAsync(httpApplication, cancellationToken);
}
//省略部分代码
}
```

省略了构造方法的代码，依然是从依赖注入中获取大量的组件并赋值给自身的成员变量，这里着重看 StartAsync 方法。其中 ASP.NET Core 的重要元素之一是 Application，也就是请求处理管道。在第 7 章中，用户的请求经过 Server 监听，处理成 HttpContex，再将这个 HttpContex 交给 Application 处理，这个处理工作就是由请求处理管道来完成的。它的类型是 RequestDelegate：

```
public delegate Task RequestDelegate(HttpContext context)
```

它由多个中间件嵌套组成，此处用到的 options.ConfigureApplication 是处理 Startup 时存储的 Startup 的 Configure 方法，这个 Configure 方法就是 ASP.NET Core 提供给我们用于自定义请求处理管道的方法。关于这部分将在第 14 章中详细介绍。

这里利用请求处理管道 application 和 HttpContextFactory 来生成一个 httpApplication，并调用 Server 的 StartAsync 方法完成应用的启动工作，httpApplication 是这个方法的重要参数。

至此，应用启动完毕，接下来等待 Server 来监听请求，并将请求传递给 HostingApplication 处理。

第9章 后台服务

大部分程序中需要用到后台服务，例如定时更新缓存或更新状态。

9.1 应用场景

以调用微信公众号的 API 为例，经常会用到 access_token，微信官方文档这样描述它：

access_token 是公众号的全局唯一接口调用凭据，有效期目前为 2 小时，需定时刷新，重复获取将导致上次获取的 access_token 失效，建议公众号开发者使用中控服务器统一获取和刷新 access_token，其他业务逻辑服务器所使用的 access_token 均来自该中控服务器，不应该各自刷新，否则容易造成冲突，导致 access_token 覆盖而影响业务。

在这个场景中，我们可以创建一个后台运行的服务，按照 access_token 的有效期定时执行请求获取新的 access_token 并存储，其他所有需要用到这个 access_token 的服务都调用这个共有的 access_token。

9.2 实现方式

ASP.NET Core 提供了一个名为 IHostedService 的接口用于自定义后台服务。IHostedService 的代码如下：

```
public interface IHostedService
{
    Task StartAsync(CancellationToken cancellationToken);
    Task StopAsync(CancellationToken cancellationToken);
}
```

通过名字可以看出，一个方法是实现这个服务启动时做的操作，另一个方法则是实现停止时做的操作。

在上一章的应用启动部分，Host 的 StartAsync 方法中有这样一段代码：

```
foreach (var hostedService in _hostedServices)
```

```
    {
        await hostedService.StartAsync(combinedCancellationToken).ConfigureAwait(false);
    }
```

作用是启动注册到系统中的 IHostedService 类型服务。默认的情况下包含了 DataProtectionHostedService 和 GenericWebHostService 两个服务。如果想自定义一个服务并运行，我们只需要做两件事：第一，实现 IHostedService 接口；第二，将这个接口实现注册到依赖注入服务中。

下面通过一个简单的例子来看实现方式。

9.2.1 实现 IHostedService 接口

新建一个类 TokenRefreshService 实现 IHostedService，如下代码所示：

```
internal class TokenRefreshService : IHostedService, IDisposable
{
    private readonly ILogger _logger;
    private Timer _timer;

    public TokenRefreshService(ILogger<TokenRefreshService> logger)
    {
        _logger = logger;
    }

    public Task StartAsync(CancellationToken cancellationToken)
    {
        _logger.LogInformation("Service starting");
        _timer = new Timer(Refresh, null, TimeSpan.Zero, TimeSpan.FromSeconds(5));
        return Task.CompletedTask;
    }

    private void Refresh(object state)
    {
        _logger.LogInformation(DateTime.Now.ToLongTimeString() + ": Refresh Token!");


        //在此写需要执行的任务
    }

    public Task StopAsync(CancellationToken cancellationToken)
    {
        _logger.LogInformation("Service stopping");
        _timer?.Change(Timeout.Infinite, 0);
        return Task.CompletedTask;
    }
```

```
    public void Dispose()
    {
        _timer?.Dispose();
    }
}
```

既然是定时刷新任务，那么要用到一个 timer，当服务启动时启动它，由它定时执行 Refresh 方法来获取新的 Token。

为了方便测试，这里规定 5 秒执行一次，实际应用时通过读取配置文件获取比较好。

9.2.2 在依赖注入中注册这个服务

在 Startup 的 ConfigureServices 中注册这个服务，如下代码所示：

```
services.AddHostedService<TokenRefreshService>();
```

运行结果如下：

```
BackService.TokenRefreshService:Information: 17:23:30: Refresh Token!
BackService.TokenRefreshService:Information: 17:23:35: Refresh Token!
BackService.TokenRefreshService:Information: 17:23:40: Refresh Token!
BackService.TokenRefreshService:Information: 17:23:45: Refresh Token!
BackService.TokenRefreshService:Information: 17:23:50: Refresh Token!
```

9.3 采用 BackgroundService 派生类的方式

在 ASP.NET Core 2.1 中提供了一个名为 BackgroundService 的类，它在 Microsoft.Extensions.Hosting 命名空间中，代码如下：

```
public abstract class BackgroundService : IHostedService, IDisposable
{
    private Task _executingTask;
    private readonly CancellationTokenSource _stoppingCts = new CancellationTokenSource();

    protected abstract Task ExecuteAsync(CancellationToken stoppingToken);
    public virtual Task StartAsync(CancellationToken cancellationToken)
    {
        _executingTask = ExecuteAsync(_stoppingCts.Token);

        if (_executingTask.IsCompleted)
        {
            return _executingTask;
        }

        return Task.CompletedTask;
    }
```

```
    public virtual async Task StopAsync(CancellationToken cancellationToken)
    {
        if (_executingTask == null)
        {
            return;
        }

        try
        {
            _stoppingCts.Cancel();
        }
        finally
        {
            await Task.WhenAny(_executingTask, Task.Delay(Timeout.Infinite, cancellation
Token));
        }

    }

    public virtual void Dispose()
    {
        _stoppingCts.Cancel();
    }
}
```

可以看出它继承自“IHostedService，IDisposable”，这相当于帮我们写好了一些“通用”的逻辑，而我们只需要继承并实现它的 ExecuteAsync 方法即可。

也就是说，我们只需在这个方法内写下这个服务需要做的事，这样刷新 Token 的 Service 就可以改写成如下代码：

```
internal class TokenRefreshService : BackgroundService
{
    private readonly ILogger _logger;
    public TokenRefreshService(ILogger<TokenRefreshService> logger)
    {
        _logger = logger;
    }
    protected override async Task ExecuteAsync(CancellationToken stoppingToken)
    {
        _logger.LogInformation("Service starting");
        while (!stoppingToken.IsCancellationRequested)
        {
            _logger.LogInformation(DateTime.Now.ToLongTimeString() + ": Refresh Token!");
            //在此写需要执行的任务
            await Task.Delay(5000, stoppingToken);
```

```
            }
            _logger.LogInformation("Service stopping");
        }
    }
```

看上去简单了不少，同样，为了方便测试，这里规定 5 秒执行一次。

9.4 注意事项

当项目部署在 IIS 上时，应用程序池回收，后台任务也会停止执行。

经测试：

（1）当 IIS 上部署的项目启动后，后台任务随之启动，任务执行相应的日志正常输出；

（2）手动回收对应的应用程序池，任务执行相应的日志输出停止；

（3）重新请求该网站，后台任务随之启动，任务执行相应的日志重新开始输出。

所以在这样的后台任务中不建议做一些需要定时执行的业务处理类的操作，可以做缓存刷新类的操作，因为当应用程序池回收后再次运行时，后台任务会随之启动。

第 10 章　依赖注入

本文通过一个维修工与工具库的例子来形象地描述为什么要用依赖注入，以及它的工作原理。根据这个例子类比 ASP.NET Core 中的依赖注入，从而深入了解它的使用方法、注意事项，以及回收机制等。

10.1 为什么要用依赖注入

软件设计原则中有一个依赖倒置原则（DIP），即要依赖于抽象，不要依赖于具体，高层模块不应该依赖于低层模块，二者应该依赖于抽象。简单地说，是为了更好地解耦。而控制反转（IoC）就是这样的原则的实现思路之一，这个思路的实现方式之一就是依赖注入（DI）。

举例说明，如图 10-1 所示，维修工老李现在要出任务去维修，得先去找库管老张申领一个扳手。

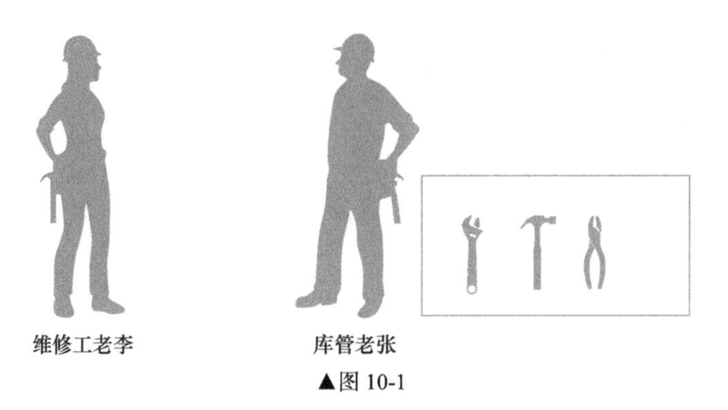

▲图 10-1

老李："请给我一把开口规格为 7mm 的六角螺母扳手。"然后老张就从仓库里拿了一把大力牌扳手给老李。

在这个例子中，老李只要告诉库管他要一个"开口规格为 7mm 的六角螺母扳手"即可，不用关心扳手的品牌和样式，也不用采购扳手，更不用关心这个扳手是怎么来的；而对于老张，只需提供满足这样规则的一个扳手即可，不用去关心老李拿着这个扳手之后去干什么，所以老李和老张都只关心“开口规格为 7mm 的六角螺母扳手”这个规则即可。也就是说，如果后期仓库里不再提供大力牌扳手，而是提供这样规则的大牛牌扳手，无论换了什么牌子和样式，只要仍满足这个规则，老李就可以正常工作。他们定义了一个规则（如接口 IWrench7mm），二者都依赖于这个规则，仓库无论提供大力牌（WrenchDaLi：IWrench7mm）还是大牛牌（WrenchDaNiu：IWrench7mm），都不影响正常工作。

这就是依赖倒置原则，不依赖于具体（牌子），高层模块（老李）不依赖于低层模块（大力牌扳手），二者应该依赖于抽象（IWrench7mm：开口规格为 7mm 的六角螺母扳手）。如果直接由老李去获取大力牌扳手，那么当业务改变要求采用大牛牌扳手时，我们要修改“老李的代码”，在本例中，我们只要在配置中让仓库由原来的提供大力牌扳手改为提供大牛牌扳手即可。老李要使用时，老张可以通过注入（构造器、属性、方法）的方式，将仓库提供的“扳手实例”提供给老李使用。

模拟代码如下：

```
public interface IWrench7mm
{
    void Screw();
}

public class WrenchDaLi:IWrench7mm
{
    public void Screw()
    {
        //拧螺母
    }

    public string Style()
    {
        return "单头扳手" ;
    }
}

public class WrenchDaNiu : IWrench7mm
{
    public void Screw()
    {
        //拧螺母
    }
    public string Style()
    {
        return "双头扳手";
    }
}
```

老李维修使用扳手时：

```
public void Repair(IWrench7mm wrench)
{
    wrench.Screw();
}
```

10.2 容器的构建和规则

继续上面的例子，老张为什么会提供给老李大力牌而不是大牛牌的扳手呢？那是因为领导老唐给了他一份仓库的“物品购置及发放清单”，这个物品仓库的构建及发放规则如图 10-2 所示。

（1）当有人要 7mm 的六角扳手时，就给他一个大力牌的扳手；当再有人要 7mm 六角螺母扳手时，就再给一把。

（2）但对于现场拍照相机，每个小组只有一台，小组内所有人共用这一台。

（3）全单位只有一辆卡车，谁申请都是使用同一辆。

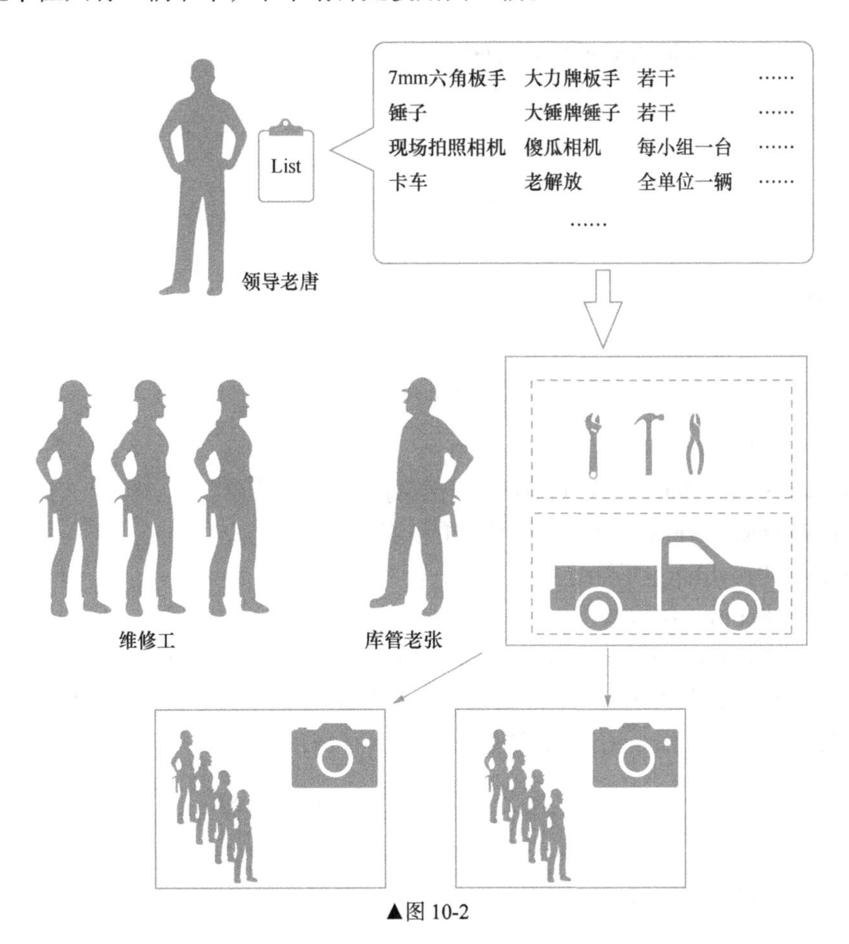

▲图 10-2

10.3 ASP.NET Core 的依赖注入

图 10-3 所示是 ASP.NET Core 中默认的依赖注入方式，看上去和图 10-2 很像。

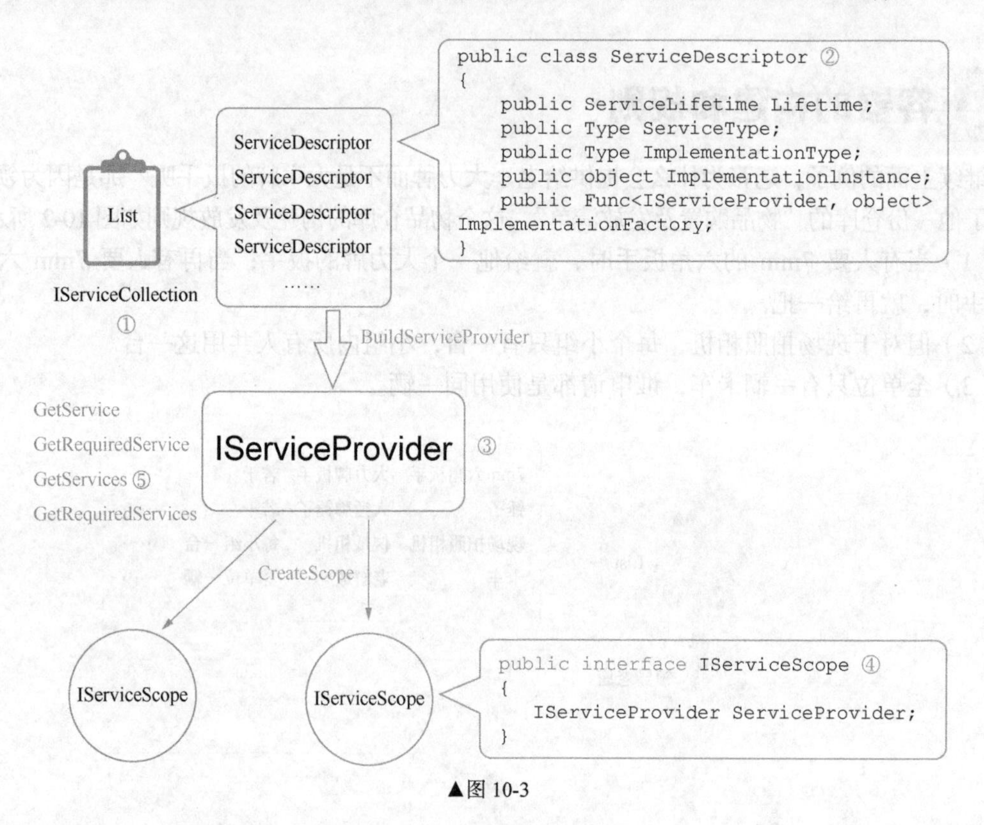

▲图 10-3

这里涉及如下关键点。

- **IServiceCollection**：领导制定的“物品购置及发放清单”。
- **ServiceDescriptor**：物品清单中的物品描述。
- **IServiceProvider**：库管老张。
- **IServiceScope**：维修小组。
- **GetService/ GetServices/ GetRequiredService/ GetRequiredServices**：借工具的方法。

前面说要将 Startup 放大来介绍，那么现在打开 Startup 这个文件，其中有 ConfigureServices 方法。顾名思义，这个方法是用来配置服务的。

```
public void ConfigureServices(IServiceCollection services)
{
    {
    services.Add Controllers With Views();
}
```

此方法接收一个 IServiceCollection 类型的参数，由其名称可以看出，它是一些服务的集合。它被定义在 Microsoft.Extensions.DependencyInjection 的 NuGet 包中，功能是依赖注入，在 ASP.NET Core 中被广泛使用。下面来介绍这几个关键的对象。

10.3.1 IServiceCollection

图 10-3 中的 IServiceCollection 就是上面例子中的“领导制定的清单”，是一个 IList<ServiceDescriptor>类型的集合。而 Startup 中的 ConfigureServices 方法的作用是让我们作为“领导”来制定这个清单。

方法中默认调用的 services.AddMvc()是调用了 IServiceCollection 的一个扩展方法 public static IMvcBuilder AddMvc(this IServiceCollection services)，其作用是向这个清单中添加一些 MVC 需要的服务，例如 Authorization、RazorViewEngin、DataAnnotations 等。

系统需要的已经添加好了，剩下的就是我们把自己需要用的添加进去。这里我们可以创建一个 ServiceDescriptor，然后把它添加到这个集合里，IServiceCollection 也提供了 AddSingleton、AddScoped 和 AddTransient 方法，3 种方法定义了所添加服务的生命周期，具体见 ServiceDescriptor。如上面的维修工例子，将大力牌扳手加入清单：

```
public void ConfigureServices(IServiceCollection services)
{
    //省略部分代码
    services.AddTransient<IWrench7mm,WrenchDaLi>();
}
```

我们可以在 ConfigureServices 中通过一系列的 Add*XXX* 将服务添加到 IServiceCollection，但实际项目中要添加很多服务，导致很多代码堆积在 ConfigureServices 方法中，不易于修改和阅读，所以推荐使用系统默认的 Add Controllers With Views()封装到一个扩展方法中。

修改上面的例子，先定义一个扩展方法：

```
    public static class MvcServiceCollectionAddWrenchExtensions
    {
        public static IServiceCollection AddWrench(this IServiceCollection services)
        {
            if (services == null)
            {
                throw new ArgumentNullException(nameof(services));
            }
            return services.AddTransient<IWrench7mm, WrenchDaLi>();
        }
    }
```

这里只是举一个简单的例子，实际在这个方法中能做很多事情。定义好方法后，在 Startup 中调用它：

```
public void ConfigureServices(IServiceCollection services)
{
    //省略部分代码
//services.AddTransient<IWrench7mm,WrenchDaLi>();
services.AddWrench();
}
```

下面来看清单中的内容。

10.3.2 ServiceDescriptor

既然 IServiceCollection 是一个 IList<ServiceDescriptor>类型的集合，那么 ServiceDescriptor 就是这个集合中的子项了，也就是“清单中物品的描述”。对照图 10-3 中的 ServiceDescriptor，它的各个属性如下。

- **Type ServiceType**：服务的类型（7mm 六角扳手）。
- **Type ImplementationType**：实现的类型（大力牌扳手）。
- **ServiceLifetime Lifetime**：服务的生命周期（谁要都给一把新的）。
- **object ImplementationInstance**：实现服务实例。
- **Func<IServiceProvider, object> ImplementationFactory**：创建服务实例的工厂。

其中 ServiceLifetime Lifetime 是一个枚举，上文中的 AddSingleton、AddScoped 和 AddTransient 就是对应这个枚举中的单例，分别如下。

- **Singleton**：单例，例子中的卡车，全单位只有一辆，谁调用都需要返回这个实例。
- **Scoped**：一定区域内的单例，例子中的傻瓜相机，每小组一台，小组内谁要都是同一台，不同小组的相机不同。
- **Transient**：临时的单例，例子中的扳手和锤子，谁要都给一把新的，所有人的都不是同一把。

从这些属性的介绍来看，ServiceDescriptor 规定了当有人需要 ServiceType 这个类型服务时，提供给它一个 ImplementationType 类型的实例，其他属性规定了提供服务的方法和生命周期。

10.3.3 IServiceProvider

IServiceProvider 服务提供者，即例子中的“老张”。由 IServiceCollection 的扩展方法 BuildServiceProvider 创建，当需要它提供某个服务时，会根据创建它的 IServiceCollection 中对应的 ServiceDescriptor 提供相应的服务实例。它提供了 GetService、GetRequiredService、GetServices、GetRequiredServices 等用于提供服务实例的方法。IserviceProvider 的作用就像老张一样，说明你需要什么服务的实例，并根据清单物品规格给你对应的工具。

GetService 和 GetRequiredService 的区别如下。

- 老李：“老张，给我一架空客 A380。”——GetService<IA380>();

- 老张："这个没有。" ——return null;
- 老李："老张，必须给我一架空客 A380！" ——GetRequiredService<IA380>();
- 老张："这个真的没有！" ——System.InvalidOperationException: "No service for type 'IA380' has been registered."；

GetServices 和 GetRequiredServices 这两个加了"s"的方法返回对应的集合。

10.3.4 IServiceScope

上文中的 ServiceDescriptor 的 Lifetime 为 Scoped 时，IServiceProvider 会为其创建一个新的接口 IServiceScope，示例代码如下：

```
public interface IServiceScope : IDisposable
{
   IServiceProvider ServiceProvider { get; }
}
```

从上面的代码可以看出，它只是对 IServiceProvider 进行了一个简单的封装。原始的 IServiceProvider 通过 CreateScope()创建了一个 IServiceScope，而这个 IServiceScope 的 ServiceProvider 属性将负责提供这个接口的服务，Lifetime 为 Scoped 的 ServiceDescriptor 创建的实例在本区域内是以"单例"的形式存在的。

在 ASP.NET Core 中，Lifetime 为 Scoped 的实例在每次请求中只创建一次。

10.4 实例获取方法及需要注意的问题

相比上面的维修工借扳手的例子，ASP.NET Core 的依赖注入有一些不同之处，如用卡车（全单位只有一辆，谁借都是这一辆）来类比单例，问题不大，但如果 A 把它借走了，B 只有等它被还回来才能去借。同样，标记为 Scoped 的傻瓜相机即使在小组内也是需要轮换使用的，这就是并发问题。对于 ASP.NET Core 的依赖注入提供的 Singleton 和 Scoped 的实例来说，它是很有可能同时被多个地方获取并调用的。

依然以维修工借扳手为例。为了方便比较是不是同一个实例，下面来修改 IWrench7mm 和 WrenchDaLi，添加一个 Guid 属性，并在构造方法里对它的 Guid 属性赋一个新值。

```
    public interface IWrench7mm
    {
        Guid Guid { get; }
        void Screw();
    }
    public class WrenchDaLi:IWrench7mm
{
        public WrenchDaLi()
```

```
    {
        this.Guid = Guid.NewGuid();
    }
    public Guid Guid { get; }
    public string Name{ get; set;}
    public void Screw()
    {
        //拧螺母
    }
    public string Style()
    {
        return "单头扳手" ;
    }
}
```

ASP.NET Core 提供了 4 种获取方法，依然以借扳手为例，创建一个 WrenchController：

```
public class WrenchController : Controller
{
    private IWrench7mm wrench;
    public WrenchController(IWrench7mm _wrench)//方法 A
    {
        wrench = _wrench;
    }

    // GET: /<controller>/
    public IActionResult Index([FromServices]IWrench7mm wrenchFromAction)//方法 B
    {
        IWrench7mm wrenchFromRequest = HttpContext.RequestServices.GetService<IWrench7mm>
(); //方法 C
        ViewBag.wrench = wrench;
        ViewBag.wrenchFromAction = wrenchFromAction;
        ViewBag.wrenchFromRequest = wrenchFromRequest;
        return View();
    }
}
```

对默认的 Action（Index）添加一个 View：

```
@inject IWrench7mm  wrenchFromView                  //方法 D
<ul>
    <li>@ViewBag.wrench.Guid</li>
    <li>@ViewBag.wrenchFromAction.Guid</li>
    <li>@ViewBag.wrenchFromRequest.Guid</li>
    <li>@wrenchFromView.Guid</li>
</ul>
```

上面的代码中，通过 4 种方法获取了 4 个 IWrench7mm 的实例。

- **在构造方法中获取：**即 wrench，在参数中直接获取已注册的 IWrench7mm 即可。
- **在 Action 中获取：**即 wrenchFromAction，类似在构造方法的参数中获取，只是这里的参数要添加[FromServices]标识。
- **在 Request 中获取：**即 wrenchFromRequest，从 HttpContext.RequestServices 中获取，注意需要使用 Microsoft.Extensions.DependencyInjection。
- **在 View 中获取：**即 wrenchFromView，可以在 View 中直接获取并使用。

也可以利用这个例子比较 AddTransient、AddSingleton、AddScoped 这 3 个方法的不同之处。

- **services.AddTransient<IWrench7mm, WrenchDaLi>()：**

```
fcd3d6af-6149-4f5d-a0da-c7053aeffa6b
5c335a8e-e129-4360-8fa2-15b0772f792f
28d33792-afef-43be-b124-07d749fd0c26
1761291c-0df6-4ebb-9b76-183420aa2e2b
```

结论：4 组值相同，说明 4 种方式获取了 4 个不同的实例。刷新页面，又变成了其他 4 个不同的值，说明即使在同一次请求中，多次获取的实例依然不同。

- **services. AddSingleton<IWrench7mm, WrenchDaLi>()：**

```
dd4c952e-b64c-4dc8-af01-2b9d667cf190
dd4c952e-b64c-4dc8-af01-2b9d667cf190
dd4c952e-b64c-4dc8-af01-2b9d667cf190
dd4c952e-b64c-4dc8-af01-2b9d667cf190
```

结论：4 组值相同，说明获得的是同一个实例。再次刷新，仍然是这 4 组值，说明无论是同一次还是多次请求，获得的结果也是同一个实例。

- **services. AddScoped<IWrench7mm, WrenchDaLi>()：**

```
ad5a600b-75fb-43c0-aee9-e90231fd510c
ad5a600b-75fb-43c0-aee9-e90231fd510c
ad5a600b-75fb-43c0-aee9-e90231fd510c
ad5a600b-75fb-43c0-aee9-e90231fd510c
```

结论：4 组值相同，但与上一次的结果不同。再次刷新，会生成其他 4 组一样的值，说明在同一次请求里，获取的实例是同一个，多次请求之间相比较，获取的实例不同。

因为无论是在 Singleton 还是 Scoped 的情况下，可能在应用的多个地方同时使用同一个实例。所以在程序设置时要注意，如果存在像上面的 WrenchDaLi 有 Name 属性提供了{ get; set; }的情况，多个引用者处理它的值，会造成一些不可预料的错误。

10.5 服务的 Dispose

对于每次请求，我们最初配置的根目录 IServiceProvider 通过 CreateScope()创建了一个新的 IServiceScope，而这个 IServiceScope 的 ServiceProvider 属性将负责该次请求的服务提供。当请求结束，这个 ServiceProvider 的 Dispose 会被调用，并负责由它创建的各个服务。

在 ASP.NET Core 1.0 中，ServiceProvider 将对所有 IDisposable 对象调用 Dispose，包括那些并非由它创建的 Dispose。

而在 ASP.NET Core 2.0 之后，ServiceProvider 只调用由它创建的 IDisposable 类型的 Dispose。例如：

```
services.AddTransient<IWrench7mm,WrenchDaLi>();
 services.AddScoped<IWrench7mm, WrenchDaNiu>();
```

这也包括通过工厂方式注册。因为即使提供了工厂，也是由容器创建的，容器会对其负责，进行自动处理。例如：

```
services.AddScoped<IWrench7mm>(t=> new WrenchDaLi());
```

但如果将一个实例添加到容器，它将不会被释放。例如：

```
services.AddSingleton<IWrench7mm>(new WrenchDaLi());
```

再次修改 WrenchDaLi 和 WrenchDaNiu，使其实现 Idisposable。举例说明，这里只需简要地通过例子看效果，并未做真正的清理工作。代码如下：

```
    public class WrenchDaLi : IWrench7mm , IDisposable
    {
        public void Dispose()
        {
            Debug.WriteLine("WrenchDaLi was disposed.");
        }

            //省略其他方法
}

    public class WrenchDaNiu : IWrench7mm , IDisposable
    {
        public void Dispose()
        {
            Debug.WriteLine("WrenchDaNiu was disposed.");
        }

            //省略其他方法
    }
```

例如通过 AddTransient 或 AddScoped 注册的部分在请求之后就会被回收。

10.6 更换容器

可以将默认的容器改为其他的容器，如 Autofac。

下面安装 Autofac 相关的包。在 NuGet 中搜索安装 Autofac 和 Autofac.Extensions.DependencyInjection 即可，如图 10-4 所示。

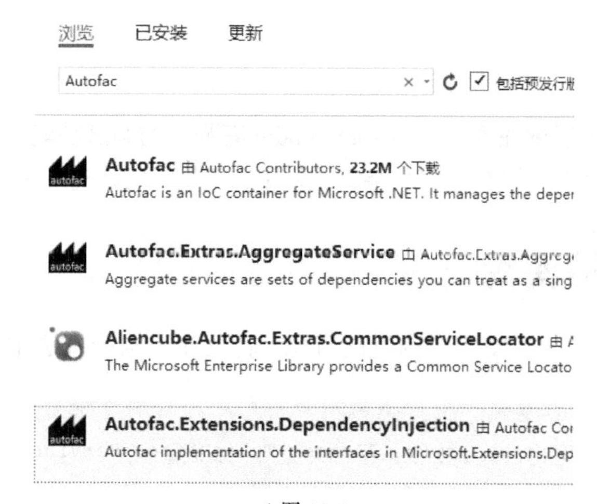

▲图 10-4

ASP.NET Core 3.0 开始采用新的更换容器方式，需要在 Program 中启用对应的工厂，见如下代码，第 3 行是新增代码：

```
public static IHostBuilder CreateHostBuilder(string[] args) =>
            Host.CreateDefaultBuilder(args)
                .UseServiceProviderFactory(new AutofacServiceProviderFactory())
                .ConfigureWebHostDefaults(webBuilder =>
                {
                    webBuilder.UseStartup<Startup>();
                });
```

相应的依赖注入注册代码写在 Startup 的 ConfigureContainer 方法中，默认情况下 Startup 中没有此方法，用户自己新增即可，示例代码如下：

```
public void ConfigureContainer(ContainerBuilder builder)
{
    builder.RegisterType<WrenchDaLi>().As<IWrench7mm>();
}
```

第 11 章 日志

应用离不开日志，虽然现在使用的 Visual Studio 有强大的调试功能，在开发过程中，对于不复杂的情况不需要输出日志，但在一些复杂的过程中，应用日常运行中的日志还是非常有用的。

11.1 内置日志的使用

第 9 章后台服务的例子中使用了内置的日志，将其注入构造方法中，直接使用即可，非常方便，简易代码如下：

```
public TokenRefreshService(ILogger<TokenRefreshService> logger)
{
    _logger = logger;
}

protected override async Task ExecuteAsync(CancellationToken stoppingToken)
{
    _logger.LogInformation("Service starting");
   //省略其他代码
}
```

运行后就可以看到输出的日志了，如图 11-1 所示。

▲图 11-1

如果想把它输出到 txt 文件中，那么试试常见的 NLog 吧。

11.2 使用 NLog 将日志输出到文件

（1）安装 Nlog：在 NuGet 中搜索并安装 NLog.Web.AspNetCore，如图 11-2 所示。

▲图 11-2

（2）添加配置文件：新建一个文件 nlog.config，并右击选择“属性”，将其“复制到输出目录”设置为“始终复制”。文件内容如下：

```
<?xml version="1.0" encoding="utf-8" ?>
<nlog xmlns="http://www.nlog-project.org/schemas/NLog.xsd"
      xmlns:xsi="http://www.w3.org/2001/XMLSchema-instance"
      autoReload="true"
      throwConfigExceptions="true"
      internalLogLevel="info"
      internalLogFile="d:\log\internal-nlog.txt">


  <!-- the targets to write to -->
  <targets>
    <!-- write logs to file  -->
    <target xsi:type="File" name="allfile" fileName="d:\log\nlog-all-${shortdate}.log"
 layout="${longdate}|${event-properties:item=EventId_Id:whenEmpty=0}|$
{uppercase:${level}}|${logger}|${message} ${exception:format=tostring}" />

    <!-- another file log, only own logs. Uses some ASP.NET core renderers -->
    <target xsi:type="File" name="ownFile-web" fileName="d:\log\nlog-own-${shortdate}.log"
layout="${longdate}|${event-properties:item=EventId_Id:whenEmpty=0}|$
{uppercase:${level}}|${logger}|${message} ${exception:format=tostring}|url: ${aspnet-
request-url}|action: ${aspnet-mvc-action}|${callsite}" />
```

```
  </targets>

  <!-- rules to map from logger name to target -->
  <rules>
    <!--All logs, including from Microsoft-->
    <logger name="*" minlevel="Trace" writeTo="allfile" />

    <!--Skip non-critical Microsoft logs and so log only own logs-->
    <logger name="Microsoft.*" maxlevel="Info" final="true" />
    <!-- BlackHole -->
    <logger name="*" minlevel="Trace" writeTo="ownFile-web" />
  </rules>
</nlog>
```

（3）修改 Program.cs 文件，在最后添加一句.UseNLog()，代码如下：

```
public static IHostBuilder CreateHostBuilder(string[] args) =>
            .ConfigureWebHostDefaults(webBuilder =>
            {
                webBuilder.UseStartup<Startup>();
            })
.UseNLog();
```

11.3 注意事项

按照 11.2 节的描述，NLog 已经可以正常使用了，但有些细节需要简要说明。

（1）文件 nlog.config 的这个名字应该是默认读取的文件名（建议全部小写，在 Linux 操作系统中要注意）。如果想使用其他名字，可以在 Program 文件中通过 ConfigureNLog 方法设置，见如下代码示例。

（2）在 11.1 节的例子中，Visual Studio 的“输出”窗口仍然在显示日志，也就是二者都处于生效状态，要想只用 NLog，可以调用 logging.ClearProviders()。

（3）注意输出的日志级别问题。

注意事项见如下代码中的注释部分：

```
public class Program
{
    public static void Main(string[] args)
    {
        //没有用默认的名字，多写了一个 1
        NLog.Web.NLogBuilder.ConfigureNLog("nlog1.config");
CreateHostBuilder(args).Build().Run();
    }
    public static IHostBuilder CreateHostBuilder(string[] args) =>
```

```
        Host.CreateDefaultBuilder(args).UseServiceProviderFactory(new AutofacService
ProviderFactory())
            .ConfigureWebHostDefaults(webBuilder =>
            {
                webBuilder.UseStartup<Startup>();
            })
            .ConfigureLogging(logging =>
            {
                //移除已经注册的其他日志处理程序
                logging.ClearProviders();
                //设置最低的日志级别
                logging.SetMinimumLevel(Microsoft.Extensions.Logging.LogLevel.Trace);
            }).UseNLog();
}
```

11.4 NLog 配置简要说明

NLog 的配置如下。

（1）日志级别：上文提到了一个日志级别，大概分为 6 个级别，由低到高排列如下。

- logger.LogTrace()。
- logger.LogDebug()。
- logger.LogInformation()。
- logger.LogWarning()。
- logger.LogError()。
- logger.LogCritical()。

（2）nlog.config 的配置规则：通过上面的例子可以看到，文件夹中输出的日志文件有 3 个，这是在 nlog.config 中配置的，通过文件名可以找到对应的配置。

- internal-nlog：记录了 NLog 的启动及加载 config 的信息。
- nlog-all：记录了所有日志。
- nlog-own：记录了自定义的日志。

这是如何产生的呢？rlog.config 中有两个关键标签——targets 和 rules。

- targets：用于配置输出相关内容，如 type 属性的可选项 File、Mail、Console 等，用于设置输出信息；layout 属性用于设置输出信息的组成元素及格式。
- rules：这个标签容易被简单理解成“规则”，而恰好例子中的 rule 正好对应了前面的两个 target，writeTo 属性指定了对应的 target。可仔细一看，两个 rule 的配置差不多，为什么下面的一个只输出了我们自定义的日志呢？由帮助文档可知，这是一个类似“路由表”的配置，日志是从上到下匹配的。<logger name="Microsoft.*" maxlevel="Info" final="true" />的 final="true" 过滤掉了“Microsoft.*”的日志。

第12章　应用的配置

大多数应用离不开配置，本章将介绍 ASP.NET Core 中常见的配置方式及系统内部的处理机制。

12.1 常见的配置方式

说到配置，我们的第一印象可能是 CONFIG 格式的 XML 文件或者 INI 格式的 INI 文件。在 ASP.NET Core 中，常用的配置文件格式为 JSON，如项目根目录中的 appsettings.json 和 appsettings.Development.json 两个文件。实际上，ASP.NET Core 支持多种配置方式，除了采用 JSON 文件的方式外，还支持内存对象、命令行等方式。

12.1.1　文件方式

文件方式是最常见的方式，ASP.NET Core 支持多种格式的配置文件，例如常见的 JSON、XML、INI 等格式的文件。

项目默认创建的配置文件 appsettings.json，其内容默认如下：

```
{
  "Logging": {
    "LogLevel": {
      "Default": "Warning"
    }
  },
  "AllowedHosts": "*"
}
```

这个文件会在系统启动时自动被加载（加载写在 Program 文件的 CreateWebHostBuilder 方法中，下一节会详细说明），默认内容主要是对日志的配置。

举个例子，如果需要在配置文件中设置应用的主题，例如颜色、风格等，那么可以在文件末尾添加如下内容：

```
  "Theme": {
    "Name": "Blue",
    "Color": "#0921DC"
  }
```

上述代码可以设置系统的主题和对应的色值。那么这个值是如何被获取并使用的呢？以默认的 HomeController 为例，新建一个名为“GetConfiguration”的 Action 来演示 Configuration 值的获取。代码如下：

```
private readonly IConfiguration _configuration;
public HomeController(IConfiguration configuration)
{
    _configuration = configuration;
}

public ContentResult GetConfiguration()
{
    return new ContentResult() { Content = $"Theme Name:{ _configuration["Theme:Name"] },
Color:{_configuration["Theme:Color"]}" };
}
```

在构造方法中通过依赖注入的方式获取了一个 IConfiguration，并在 Action 中通过这个 IConfiguration 获取了 appsettings.json 中设置的值。可以看出，在获取值时，是通过“:”符号来体现 JSON 文件的层级关系的，例如获取“Color”的值，对应的表达式为“_configuration["Theme:Color"]”。这是因为整个 JSON 文件会被处理为 Key-Value 格式，如本例中的“Theme”的两个值会被分解为如表 12-1 所示的格式。

表 12-1

Key	Value
Theme:Name	Blue
Theme:Color	#0921DC

这里有两点说明：第一，“Key”不区分大小写，即与“theme:color”是等效的；第二，默认“Value”值是 String 类型。

除了上例中的获取方式，还可以通过 GetValue 方法获取，代码如下。

```
_configuration.GetValue<string>("Theme:Color","#000000")
```

意思是将获取到的值转换为 string 类型，如果获取失败，则返回默认值“#000000”。

本例演示了系统默认的 appsettings.json 中的内容如何被自动加载，那么如何将自定义的 JSON 文件中的内容应用到系统的配置中呢？

新建一个名为“Theme.json”的文件，再预设一个红色主题，代码如下：

```
{
  "Theme": {
```

```
    "Name": "Red",
    "Color": "#FF4500"
  }
}
```

由于这个自定义的 Theme.json 文件不会被自动加载，因此需要手动将其添加到系统的配置中。在介绍应用系统的启动时提到，配置是在 Program 文件的 CreateDefaultBuilder 方法中被加载的，可以在其后继续通过 ConfigureAppConfiguration 方法设置。例如以下代码：

```
public static IWebHostBuilder CreateWebHostBuilder(string[] args) =>
    WebHost.CreateDefaultBuilder(args)
.ConfigureAppConfiguration((hostingContext,config)=> {

        config.SetBasePath(Directory.GetCurrentDirectory());
        var path = Path.Combine(Directory.GetCurrentDirectory(), "Theme.json");
        config.AddJsonFile(path, optional: false, reloadOnChange: true);

    })
     .ConfigureWebHostDefaults(webBuilder =>
     {
         webBuilder.UseStartup<Startup>();
     });
    }
```

首先通过 SetBasePath 方法设置基本路径，然后通过 AddJsonFile 方法添加“Theme.json”文件。这个方法有 3 个参数，第一个设置“Theme.json”所在的位置，第二个设置此文件是否可选，第三个设置当此文件被修改后是否自动重新加载该文件。

再次访问 Home/GetConfiguration，返回结果如下：

```
Theme Name:Red,Color:#FF4500
```

这是因为后添加的 Theme.json 文件中的 Theme 值覆盖了 appsettings.json 中的 Theme 值。这里涉及各种配置方式的优先级问题，在下一节会讲。

在介绍了将 JSON 格式的文件用作配置的例子后，再来看如何采用 INI 格式。新建一个名为 Theme.ini 的文件，为了不覆盖之前设置的 Theme，本例将 Theme 改为了 ThemeGreen。

```
[ThemeGreen]
Name=Green
Color=#76EE00
```

通过 ConfigureAppConfiguration 方法将这个 INI 文件添加到配置中去。

```
var pathIni = Path.Combine(Directory.GetCurrentDirectory(), "Theme.ini");
config.AddIniFile(pathIni, optional: false, reloadOnChange: true);
```

修改 Action 中读取配置的 Key，将对应的 Theme 改为 ThemeGreen：

```
public ContentResult GetConfiguration()
{
    return new ContentResult() { Content = $"Theme
Name:{ _configuration["ThemeGreen:Name"] },Color:{_configuration.GetValue<string>
("ThemeGreen:Color", "#000000")}" };
}
```

再次访问 home/GetConfiguration，返回结果如下：

```
Theme Name:Green,Color:#76EE00
```

上面介绍了 JSON 和 INI 两种格式的文件的应用，除了二者文件格式的不同以及被添加到配置时采用的方法不同（分别采用了 AddJsonFile 和 AddIniFile 方法），在其他环节二者的使用方式均是一样的。同理，对于 XML 格式的文件，有一个对应的 AddXmlFile 方法可使用，而其他环节和 JSON、INI 文件的应用也是一样的，此处就不再举例描述了。

12.1.2 目录和文件

除了上一节利用 JSON、INI 和 XML 等常用的文件格式，还可以将指定目录和文件作为配置的数据来源。

例如，文件夹 s 下面有 1.txt 和 2.txt 两个文件，文件内容分别是 s1 和 s2，如图 12-1 所示。

▲图 12-1

可以将这一目录和文件作为配置的数据来源，同样只需要在 ConfigureAppConfiguration 方法中添加即可，见如下代码：

```
var pathFile = Path.Combine(Directory.GetCurrentDirectory(), "s");
config.AddKeyPerFile(directoryPath: pathFile, optional: true);
```

通过一个 Action 测试一下：

```
public ContentResult GetFileConfiguration()
{
     return new ContentResult() { Content = $"1.txt:{_configuration["1.txt"]},
       2.txt:{_configuration["2.txt"]}" };
}
```

返回结果为：

```
1.txt:s1,2.txt:s2
```

可见这样的方法是将 s 文件夹下的两个文件的文件名作为 Key，文件内容作为 Value。

12.1.3 命令行

通过命令行启动应用时，可以在命令行中通过添加 Key-Value 的方式作为配置的数据来源，例如执行如下命令启动应用：

```
dotnet run key1=value1 key2=value2
```

访问定义好的如下 Action：

```
public ContentResult GetCommandConfiguration()
{
    return new ContentResult() { Content = $"key1:{_configuration["key1"]},
          key2:{_configuration["key2"]}" };
}
```

返回结果为：

```
key1:value1,key2:value2
```

这是由于在默认的 WebHost.CreateDefaultBuilder(args)方法中添加了对命令行参数的调用的内容。如果需要在 ConfigureAppConfiguration 方法中继续添加，只需要在该方法中调用 config.AddCommandLine(args)方法。

12.1.4　环境变量

在 WebHost.CreateDefaultBuilder(args)方法中，除了会加载命令行参数，还会加载环境变量中的数据。此处的环境变量包括系统的环境变量，如图 12-2 所示。

▲图 12-2

环境变量中的“变量”和“值”会被读取为配置的 Key 和 Value。

除了读取系统的环境变量，也可以在项目的属性中添加，如图 12-3 所示。

▲图 12-3

除了我们熟悉的名为 ASPNETCORE_ENVIRONMENT 的环境变量，这里又添加了一个 Key 为 TestKey、Value 为 TestValue 的环境变量。

添加一个 Action 测试一下：

```
public ContentResult GetEnvironmentVariables()
{
    return new ContentResult() { Content = $"TestKey:{_configuration["TestKey"]}, OS:
{_configuration["OS"]}" };
}
```

分别读取图 12-2 和图 12-3 中的两个环境变量，访问这个 Action，返回结果为：

```
TestKey:TestValue,OS:Windows_NT
```

12.1.5 内存对象

以上的例子都是读取一些外部的数据源并将其转换成了配置中的 Key-Value 格式，那么是否可以直接在应用中通过代码方式创建一些 Key-Value 并加入配置呢？当然是可以的。常见的是 Dictionary，新建一个 Dictionary，代码如下：

```
public static readonly Dictionary<string, string> dict = new Dictionary<string,
string> { { "ThemeName", "Purple" },{"ThemeColor", "#7D26CD" } };
```

在 ConfigureAppConfiguration 方法中将其加入配置：

```
config.AddInMemoryCollection(dict);
```

新建一个 Action 进行测试：

```
public ContentResult GetMemoryConfiguration()
{
    return new ContentResult() { Content = "ThemeName:{_configuration["ThemeName"]},
      ThemeColor:{_configuration["ThemeColor"]}" };
}
```

返回结果为：

```
ThemeName:Purple,ThemeColor:#7D26CD
```

12.2 内部处理机制解析

上一节介绍了配置的数据源被注册、加载和获取的过程，本节来看系统是如何实现这个过程的。

12.2.1 数据源的注册

在上一节介绍的数据源设置中，appsettings.json、命令行、环境变量 3 种方式是被系统自动加载的，这是因为系统在 webHost.CreateDefaultBuilder(args)中已经为这 3 种数据源进行了注册。这个方法同样调用了 ConfigureAppConfiguration 方法，代码如下：

```
public static IWebHostBuilder CreateDefaultBuilder(string[] args)
{
    var builder = new WebHostBuilder();

    //省略部分代码

    builder.ConfigureAppConfiguration((hostingContext, config) =>
    {
        var env = hostingContext.HostingEnvironment;

        config.AddJsonFile("appsettings.json", optional: true, reloadOnChange: true)
                .AddJsonFile($"appsettings.{env.EnvironmentName}.json", optional: true,
reloadOnChange: true);

        if (env.IsDevelopment() && !string.IsNullOrEmpty(env.ApplicationName))
        {
```

```
                var appAssembly = Assembly.Load(new AssemblyName(env.ApplicationName));
                if (appAssembly != null)
                {
                    config.AddUserSecrets(appAssembly, optional: true);
                }
            }

            config.AddEnvironmentVariables();

            if (args != null)
            {
                config.AddCommandLine(args);
            }
    })

        //省略部分代码

        return builder;
    }
```

ConfigureAppConfiguration 方法加载的内容主要有 4 种。第一种加载的是 appsettings.json 和 appsettings.{env.EnvironmentName}.json 两个 JSON 文件，关于 env.EnvironmentName 在前面的章节已经说过，常见的有 Development、Staging 和 Production 这 3 种值，我们一般在开发调试时采用第一种 Development，也就是会加载 appsettings.json 和 appsettings. Development.json 两个 JSON 文件。第二种加载的是用户机密文件，这仅限于 Development 状态下，通过 config.AddUserSecrets 方法加载。第三种是通过 config.AddEnvironmentVariables 方法加载的环境变量。第四种是通过 config.AddCommandLine 方法加载的命令行参数。

注意：这里的 ConfigureAppConfiguration 方法此时是不会被执行的，只是将这个方法作为一个 Action<WebHostBuilderContext, IConfigurationBuilder> configureDelegate 合并到 WebHostBuilder 的 _configureAppConfigurationBuilder 属性中。_configureAppConfigurationBuilder 是一个 Action <WebHostBuilderContext, IConfigurationBuilder>类型的 Action。对应的代码如下：

```
public IWebHostBuilder ConfigureAppConfiguration(Action<WebHostBuilderContext,
IConfigurationBuilder> configureDelegate)
{
    _configureAppConfigurationBuilder += configureDelegate;
    return this;
}
```

在上一节的例子中，我们在 webHost.CreateDefaultBuilder(args)方法之后再次调用 Configure AppConfiguration 方法，并添加了一些自定义的数据源，这个方法也没有执行，同样被合并到了名为_configureAppConfigurationBuilder 的 Action 中。直到 WebHostBuilder 通过它的 Build

方法创建 WebHost 时，才会执行这个 Action。这段代码写在被 Build 方法调用的 BuildCommonServices()中：

```
private IServiceCollection BuildCommonServices(out AggregateException
hostingStartupErrors)
{
    //省略部分代码

    var builder = new ConfigurationBuilder()
        .SetBasePath(_hostingEnvironment.ContentRootPath)
        .AddConfiguration(_config, shouldDisposeConfiguration: true);

    _configureAppConfigurationBuilder?.Invoke(_context, builder);

    var configuration = builder.Build();
    services.AddSingleton<IConfiguration>(_ => configuration);
    _context.Configuration = configuration;

    //省略部分代码

    return services;
}
```

那么在_configureAppConfigurationBuilder 被执行时，这些不同的数据源是如何被加载的呢？这部分功能在 namespace Microsoft.Extensions.Configuration 命名空间中实现。

以 appsettings.json 对应的 config.AddJsonFile("appsettings.json", optional: true, reloadOnChange: true)方法为例，我们来看它的实现方式。IConfigurationBuilder 接口对应的实现是 ConfigurationBuilder，代码如下：

```
public class ConfigurationBuilder : IConfigurationBuilder
    {
        public IList<IConfigurationSource> Sources { get; } = new List<IConfiguration
Source>();

        public IDictionary<string, object> Properties { get; } = new Dictionary<string,
object>();

        public IConfigurationBuilder Add(IConfigurationSource source)
        {
            if (source == null)
            {
                throw new ArgumentNullException(nameof(source));
            }

            Sources.Add(source);
```

```
            return this;
        }
        //省略了 IConfigurationRoot Build 方法，下文介绍
    }
```

ConfigureAppConfiguration 方法中调用的 AddJsonFile 方法来自 JsonConfigurationExtensions 类，代码如下：

```
public static class JsonConfigurationExtensions
{
    //省略部分代码

    public static IConfigurationBuilder AddJsonFile(this IConfigurationBuilder builder,
 IFileProvider provider, string path, bool optional, bool reloadOnChange)
    {
        if (builder == null)
        {
            throw new ArgumentNullException(nameof(builder));
        }
        if (string.IsNullOrEmpty(path))
        {
            throw new ArgumentException(Resources.Error_InvalidFilePath, nameof(path));
        }

        return builder.AddJsonFile(s =>
        {
            s.FileProvider = provider;
            s.Path = path;
            s.Optional = optional;
            s.ReloadOnChange = reloadOnChange;
            s.ResolveFileProvider();
        });
    }

    public static IConfigurationBuilder AddJsonFile(this IConfigurationBuilder builder,
 Action<JsonConfigurationSource> configureSource)
        => builder.Add(configureSource);
}
```

AddJsonFile 方法会创建一个 JsonConfigurationSource，并通过 ConfigurationBuilder 的 Add(IConfigurationSource source)方法将 JsonConfigurationSource 添加到 ConfigurationBuilder 的 IList<IConfigurationSource> Sources 集合中。

同理，针对环境变量，存在对应的 EnvironmentVariablesExtensions，会创建一个对应的 EnvironmentVariablesConfigurationSource 并添加到 ConfigurationBuilder 的 IList<IConfigurationSource> Sources 集合中。类似地，还有 CommandLineConfigurationExtensions 和 CommandLineConfiguration

Source 等。最终结果是根据数据源的加载顺序，生成多个 *XXX*ConfigurationSource 对象（都直接或间接实现了 IConfigurationSource 接口）添加到 ConfigurationBuilder 的 IList<IConfigurationSource> Sources 集合中。

Program 文件的 WebHost.CreateDefaultBuilder(args)方法中的 ConfigureAppConfiguration 方法被调用后，如果在 CreateDefaultBuilder 方法之后再次调用了 ConfigureAppConfiguration 方法并添加了数据源（如上一节的例子），同样会生成相应的 *XXX*ConfigurationSource 对象添加到 ConfigurationBuilder 的 IList<IConfigurationSource> Sources 集合中。

注意：这里不是每一种数据源都生成一个 *XXX*ConfigurationSource，而是根据每次添加生成一个 *XXX*ConfigurationSource，并且遵循添加的先后顺序。例如添加多个 JSON 文件，会生成多个 JsonConfigurationSource。

这些 ConfigurationSource 之间的关系如图 12-4 所示。

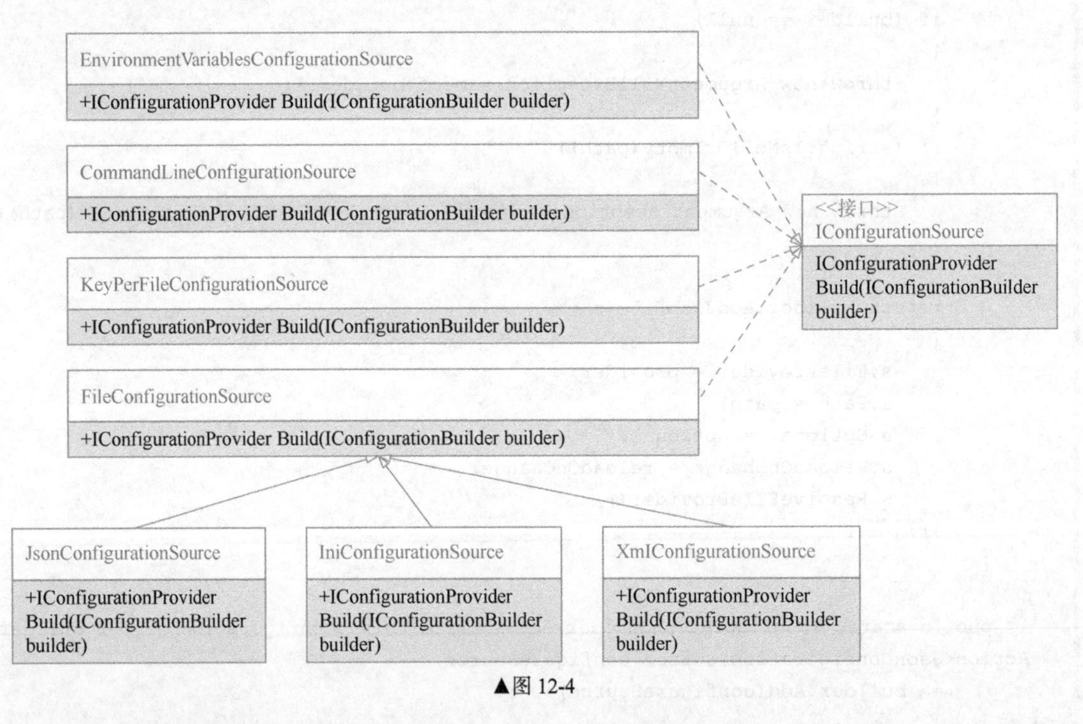

▲图 12-4

至此各种数据源的收集工作就完成了，并且都添加到了 ConfigurationBuilder 的 IList<IConfigurationSource> Sources 集合中。

回到 BuildCommonServices 方法中，通过 foreach 循环逐一执行 configureAppConfiguration 方法，获取 IList<IConfigurationSource>后，接着 var configuration = builder.Build()，调用 ConfigurationBuilder 的 Build 方法创建了一个 IConfigurationRoot 对象。对应的代码如下：

```
public class ConfigurationBuilder : IConfigurationBuilder
    {
```

```
        public IList<IConfigurationSource> Sources { get; } = new List
<IConfigurationSource>();

        //省略部分代码

        public IConfigurationRoot Build()
        {
            var providers = new List<IConfigurationProvider>();
            foreach (var source in Sources)
            {
                var provider = source.Build(this);
                providers.Add(provider);
            }
            return new ConfigurationRoot(providers);
        }

    }
```

这个方法主要体现了两个过程：第一，遍历 IList<IConfigurationSource> Sources 集合，主要调用其中的各个 IConfigurationSource 的 Build 方法创建对应的 IConfigurationProvider，最终生成一个 List<IConfigurationProvider>；第二，通过集合 List<IConfigurationProvider>创建 ConfigurationRoot，ConfigurationRoot 实现 IConfigurationRoot 接口。

对于第一个过程，依然以 JsonConfigurationSource 为例，代码如下：

```
    public class JsonConfigurationSource : FileConfigurationSource
    {
        public override IConfigurationProvider Build(IConfigurationBuilder builder)
        {
            EnsureDefaults(builder);
            return new JsonConfigurationProvider(this);
        }
}
```

JsonConfigurationSource 会通过 Build 方法创建一个名为 JsonConfigurationProvider 的对象。由 JsonConfigurationProvider 的名字可知，它是针对 JSON 文件的，这就意味着不同类型的 IConfigurationSource 创建的 IConfigurationProvider 类型是不一样的，对应图 12-4 中的 IConfigurationSource，生成的 IConfigurationProvider 关系如图 12-5 所示。

系统中添加的多个数据源被转换成了一个个对应的 ConfigurationProvider，这些 ConfigurationProvider 组成了一个 ConfigurationProvider 的集合。

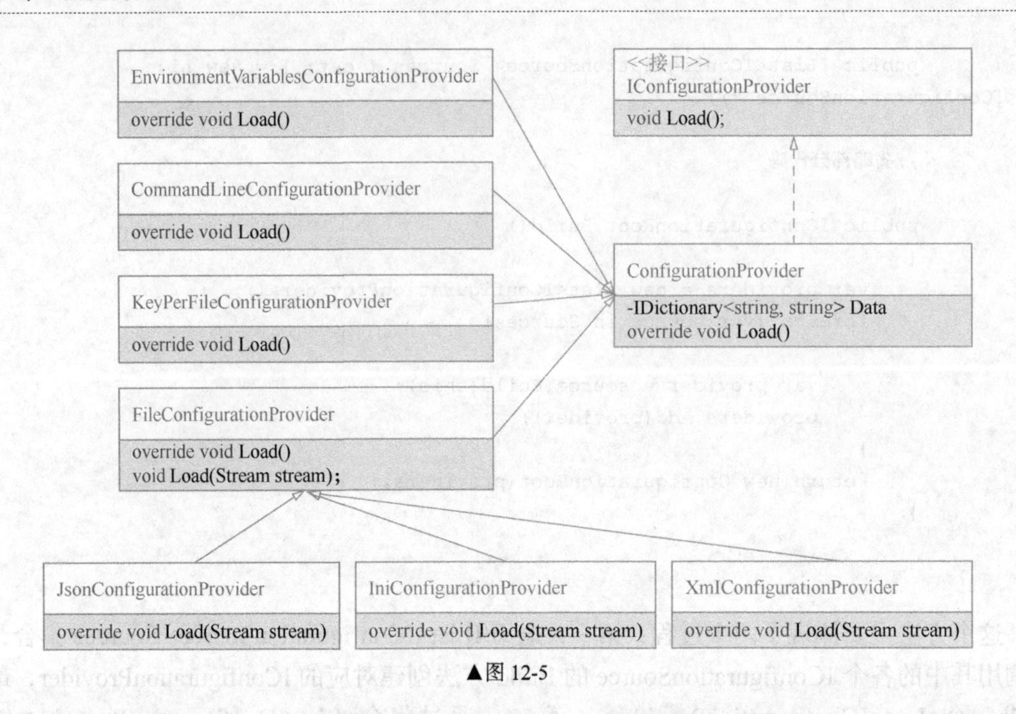

▲图 12-5

对于第二个过程，ConfigurationBuilder 的 Build 方法的最后一句是 return new ConfigurationRoot(providers)，即通过第一个过程创建的 ConfigurationProvider 的集合创建 ConfigurationRoot。代码如下：

```
public class ConfigurationRoot : IConfigurationRoot, IDisposable
    {
        private readonly IList<IConfigurationProvider> _providers;
        private readonly IList<IDisposable> _changeTokenRegistrations;
        private ConfigurationReloadToken _changeToken = new ConfigurationReloadToken();

        public ConfigurationRoot(IList<IConfigurationProvider> providers)
        {
            //省略验证代码
            _providers = providers;
            _changeTokenRegistrations = new List<IDisposable>(providers.Count);
            foreach (var p in providers)
            {
                p.Load();
                _changeTokenRegistrations.Add(ChangeToken.OnChange(() =>
p.GetReloadToken(), () => RaiseChanged()));
            }
        }
        //省略部分代码
}
```

可以看出，ConfigurationRoot 的构造方法的主要作用是将 ConfigurationProvider 的集合作

为自己的一个属性的值，并遍历这个集合，逐一调用这些 ConfigurationProvider 的 Load 方法，并为 ChangeToken 的 OnChange 方法绑定数据源的改变通知和处理方法。

12.2.2 数据源的加载

由图 12-5 可知，所有类型的数据源最终创建的 *XXX*ConfigurationProvider 都继承自 ConfigurationProvider，所以它们都有一个 Load 方法和一个 IDictionary<string, string> 类型的 Data 属性，这是整个配置系统的重要核心。Load 方法用于数据源的数据的读取与处理，而 Data 用于保存最终结果，通过逐一调用 Provider 的 Load 方法完成了整个配置系统的数据加载。

以 JsonConfigurationProvider 为例，它继承自 FileConfigurationProvider，FileConfigurationProvider 的代码如下：

```
public abstract class FileConfigurationProvider : ConfigurationProvider, IDisposable
{
    //省略部分代码
    private void Load(bool reload)
    {
        var file = Source.FileProvider?.GetFileInfo(Source.Path);
        if (file == null || !file.Exists)
        {
            //省略部分代码
        }
        else
        {
            if (reload)
            {
                Data = new Dictionary<string, string>(StringComparer.OrdinalIgnoreCase);
            }
            using (var stream = file.CreateReadStream())
            {
                try
                {
                    Load(stream);
                }
                catch (Exception e)
                {
                  HandleException(ExceptionDispatchInfo.Capture(e));
                  }
            }
        }
        OnReload();
    }
    public override void Load()
    {
        Load(reload: false);
    }
```

```
    public abstract void Load(Stream stream);
  }
```

本段代码的主要功能是读取文件生成 stream，然后调用 Load(stream)方法解析文件内容。由图 12-5 可知，JsonConfigurationProvider、IniConfigurationProvider、XmlConfigurationProvider 都继承自 FileConfigurationProvider。而对于 JSON、INI、XML 这 3 种数据源来说，只是文件的格式不同，所以将通用的读取文件内容的功能交给了 FileConfigurationProvider 来完成，而这 3 个子类的 ConfigurationProvider 只需要将 FileConfigurationProvider 读取到的文件内容解析即可。所以这个参数为 stream 的 Load 方法写在 JsonConfigurationProvider、IniConfigurationProvider、XmlConfigurationProvider 等子类中，用于专门处理自身对应的格式的文件。

JsonConfigurationProvider 的代码如下：

```
public class JsonConfigurationProvider : FileConfigurationProvider
{
    public JsonConfigurationProvider(JsonConfigurationSource source) : base(source) { }

    public override void Load(Stream stream)
    {
        try
        {
            Data = JsonConfigurationFileParser.Parse(stream);
        }
        catch (JsonException e)
        {
            throw new FormatException(Resources.Error_JSONParseError, e);
        }
}
  }
```

JsonConfigurationProvider 中关于 JSON 文件的解析由 JsonConfigurationFileParser.Parse(stream)完成，最终的解析结果被赋值给父类 FileConfigurationProvider 的名为 Data 的属性。

所以最终每个数据源的内容都分别被解析成了 IDictionary<string, string>集合，这个集合作为对应的 ConfigurationProvider 的一个属性，而众多 ConfigurationProvider 组成的集合又作为 ConfigurationRoot 的属性，最终它们的关系如图 12-6 所示。

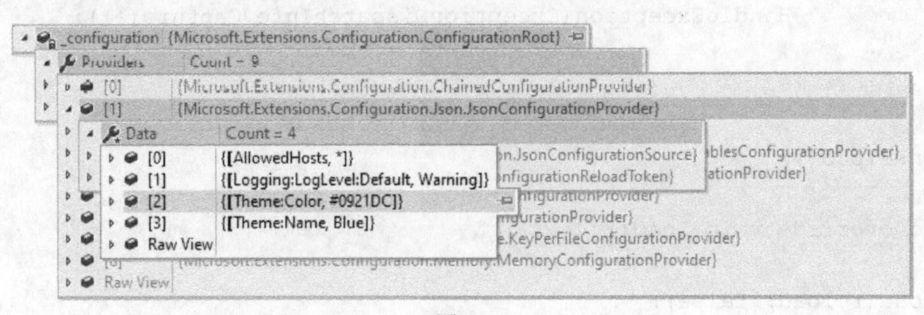

▲图 12-6

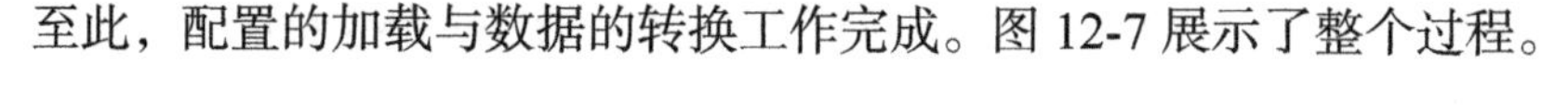

至此，配置的加载与数据的转换工作完成。图 12-7 展示了整个过程。

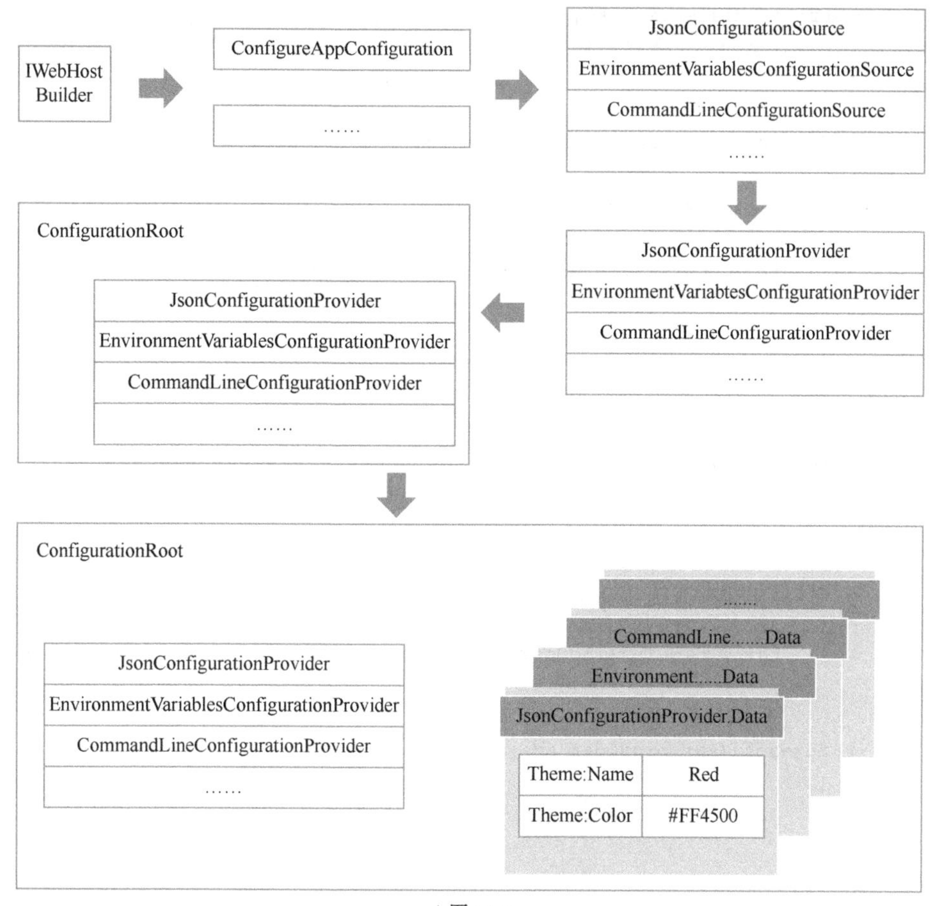

▲图 12-7

12.2.3 配置的读取

在 12.1 节中，通过_configuration["Theme:Color"]的方式获取了对应的配置值，这是如何实现的呢？现在我们已经了解了数据源的加载过程，而_configuration 就是数据源被加载后的最终产物，即 ConfigurationRoot，如图 12-7 所示。它的代码如下：

```
public class ConfigurationRoot : IConfigurationRoot, IDisposable
{
    private readonly IList<IConfigurationProvider> _providers;
    private readonly IList<IDisposable> _changeTokenRegistrations;
    private ConfigurationReloadToken _changeToken = new ConfigurationReloadToken();

    //省略上文已讲过的构造方法
```

```
    public IEnumerable<IConfigurationProvider> Providers => _providers;

    public string this[string key]
    {
        get
        {
            for (var i = _providers.Count - 1; i >= 0; i--)
            {
                var provider = _providers[i];

                if (provider.TryGet(key, out var value))
                {
                    return value;
                }
            }

            return null;
        }
        set
        {
            if (!_providers.Any())
            {
                throw new InvalidOperationException(Resources.Error_NoSources);
            }

            foreach (var provider in _providers)
            {
                provider.Set(key, value);
            }
        }
    }

    public IEnumerable<IConfigurationSection> GetChildren() =>
this.GetChildrenImplementation(null);

    public IChangeToken GetReloadToken() => _changeToken;

    public IConfigurationSection GetSection(string key)
        => new ConfigurationSection(this, key);

    public void Reload()
    {
        foreach (var provider in _providers)
        {
            provider.Load();
        }
```

```
            RaiseChanged();
        }

        private void RaiseChanged()
        {
            var previousToken = Interlocked.Exchange(ref _changeToken, new
    ConfigurationReloadToken());
            previousToken.OnReload();
        }

    //省略部分代码

    }
```

对应_configuration["Theme:Color"]的读取方式的是索引器“string this[string key]”，通过查看其 get 方法可知，它通过倒序遍历所有的 ConfigurationProvider，在 ConfigurationProvider 的 Data 中尝试查找是否存在 Key 为“Theme:Color”的值。这也说明了在 12.1 节的例子中，在 Theme.json 中设置了 Theme 对象的值后，原本在 appsettings.json 设置的 Theme 的值被覆盖的原因。从图 12-6 中可以看出，appsettings.json 设置的 Theme 值其实也是被读取并加载的，只是由于 ConfigurationRoot“倒序”遍历 ConfigurationProvider 的方式，导致后注册的 Theme.json 中的 Theme 值先被查找到了，同时验证了“所有配置值均认为是 string 类型”的约定。

ConfigurationRoot 还有一个 GetSection 方法，会返回一个 IConfigurationSection 对象，对应的是 ConfigurationSection 类，代码如下：

```
public class ConfigurationSection : IConfigurationSection
    {
        private readonly IConfigurationRoot _root;
        private readonly string _path;
        private string _key;

        public ConfigurationSection(IConfigurationRoot root, string path)
        {
            if (root == null)
            {
                throw new ArgumentNullException(nameof(root));
            }

            if (path == null)
            {
                throw new ArgumentNullException(nameof(path));
            }

            _root = root;
            _path = path;
```

```
        }

        public string Path => _path;
        public string Key
        {
            get
            {
                if (_key == null)
                {
                }
                return _key;
            }
        }
        public string Value
        {
            get
            {
                return _root[Path];
            }
            set
            {
                _root[Path] = value;
            }
        }
        public string this[string key]
        {
            get
            {
                return _root[ConfigurationPath.Combine(Path, key)];
            }

            set
            {
                _root[ConfigurationPath.Combine(Path, key)] = value;
            }
        }

        public IConfigurationSection GetSection(string key) =>
_root.GetSection(ConfigurationPath.Combine(Path, key));

        public IEnumerable<IConfigurationSection> GetChildren() =>
_root.GetChildrenImplementation(Path);

        public IChangeToken GetReloadToken() => _root.GetReloadToken();
    }
```

代码看上去很简单，可以说是没有什么实质的功能，只是保存了当前路径和对ConfigurationRoot的引用。它的方法大多是通过调用ConfigurationRoot的对应方法完成的，通过自身的路径计算在ConfigurationRoot中对应的Key，从而获取对应的值。ConfigurationRoot对配置值的读取功能以及数据源的重新加载功能（Reload方法）是通过ConfigurationProvider实现的，实际数据保存在ConfigurationProvider的Data值中。所以ConfigurationRoot和ConfigurationSection就像一个外壳，自身并不负责数据源的加载（重载）与存储，只负责实现一个配置值的读取功能。

由于配置值的读取是按照数据源加载顺序的倒序进行的，所以对于Key值相同的多个配置，只会读取后加载的数据源中的配置，因此ConfigurationRoot和ConfigurationSection就模拟出了一个树状结构，如图12-8所示。

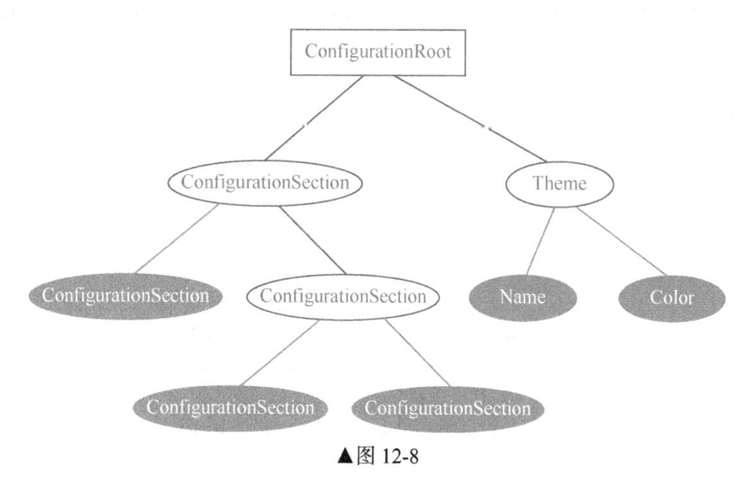

▲图12-8

图12-8以如下配置为例：

```
{
  "Theme": {
    "Name": "Blue",
    "Color": "#0921DC"
  }
}
```

ConfigurationRoot利用其制定的读取规则，将这样的配置模拟成图12-8的树状结构，它有以下特性。

（1）所有节点都被认为是一个ConfigurationSection，不同的是“Theme”这样的节点的值为空（图中用空心椭圆表示），而Name和Color这样的节点有对应的值（图中用实心椭圆表示）。

（2）由于对Key值相同的多个配置会读取后加载的数据源中的配置，所以不会出现相同路径的同名节点。例如12.1节例子中多种数据源配置了“Theme”值，在这里只会体现最后加载的配置项。

12.2.4　配置的更新

由于 ConfigurationRoot 未实际保存数据源中加载的配置值，所以配置的更新实际还是由对应的 ConfigurationProvider 来完成。以 JsonConfigurationProvider、IniConfigurationProvider、XmlConfigurationProvider 为例，它们的数据源都是具体文件，所以对文件内容的改变的监控也放在 FileConfigurationProvider 中实现。FileConfigurationProvider 的构造方法中添加了对应文件的监控设置，这里会首先判断数据源的 ReloadOnChange 参数是否被设置为 True。

```
public abstract class FileConfigurationProvider : ConfigurationProvider
{
    public FileConfigurationProvider(FileConfigurationSource source)
    {
        if (source == null)
        {
            throw new ArgumentNullException(nameof(source));
        }
        Source = source;

        if (Source.ReloadOnChange && Source.FileProvider != null)
        {
            _changeTokenRegistration = ChangeToken.OnChange(
                () => Source.FileProvider.Watch(Source.Path),
                () => {
                    Thread.Sleep(Source.ReloadDelay);
                    Load(reload: true);
                });
        }   }
    //省略其他代码
}
```

当数据源发生改变并且 ReloadOnChange 被设置为 True 时，对应的 ConfigurationProvider 就会重新加载数据，但 ConfigurationProvider 更新数据源不会改变它在 ConfigurationRoot 的 IEnumerable<IConfigurationProvider>列表中的顺序。如果列表中存在 A 和 B 两个 Configuration Provider，并且含有相同的配置项，B 排在 A 后面，那么对于这些相同的配置项，A 中的配置项是被 B 中的“覆盖”的。即使 A 的数据更新了，它依然处于“被覆盖”的位置，应用中读取相应配置项时依然是读取 B 中的配置项。

12.2.5　配置的绑定

在 12.1 节的例子中讲了两种获取配置值的方式，类似_configuration["Theme:Name"]和_configuration.GetValue<string>("Theme:Color","#000000")可以获取到 Theme 的 Name 和 Color 的值，那么就会有下面的疑问。

appsettings.json 中存在如下配置：

```
{
  "Theme": {
    "Name": "Blue",
    "Color": "#0921DC"
  }
}
```

新建一个 Theme 类如下：

```
public class Theme
{
    public string Name { get; set; }
    public string Color { get; set; }
}
```

是否可以将配置值获取并赋值到这样的一个 Theme 的实例中呢？当然可以，系统提供了这样的功能，可以采用如下代码实现：

```
Theme theme = new Theme();
_configuration.GetSection("Theme").Bind(theme);
```

绑定功能由 ConfigurationBinder 实现，其逻辑不复杂，感兴趣的读者可自行查看其代码。

第13章　配置的 Options 模式

上一章讲到了配置的用法及内部处理机制，对于配置，ASP.NET Core 还提供了一种 Options 模式。

13.1 Options 的使用

在上一章配置的绑定案例中，可以将配置绑定到一个 Theme 实例中。也就是在使用对应配置时，需要进行一次绑定操作。Options 模式提供了更直接的方式，并且可以通过依赖注入的方式提供配置的读取。下文中称每一条 Options 配置为 Option。

13.1.1　简单的不为 Option 命名的方式

依然采用上一章的例子，在 appsettings.json 中存在如下配置：

```
{
  "Theme": {
    "Name": "Blue",
    "Color": "#0921DC"
  }
}
```

修改 ValueController，代码如下：

```
public class ValuesController : Controller
{
    private IOptions<Theme> _options = null;
    public ValuesController(IOptions<Theme> options)
    {
        _options = options;
    }

    public ContentResult GetOptions()
    {
```

```
        return new ContentResult() { Content = $"options:{ _options.Value.Name}" };
    }
}
```

依然需要在 Startup 中注册：

```
public void ConfigureServices(IServiceCollection services)
{
    //省略部分代码
services.Configure<Theme>(Configuration.GetSection("Theme"));
}
```

请求 Action，获取的结果为：

```
options:Blue
```

这样就可以通过依赖注入的方式使用该配置了。这里将“Theme”绑定了这样的配置，但如果有多个这样的配置呢？例如：

```
  "Themes": [
    {
      "Name": "Blue",
      "Color": "#0921DC"
    },
    {
      "Name": "Red",
      "Color": "#FF4500"
    }
  ]
```

在这种情况下存在多个 Theme 的配置，以之前依赖注入的方式配置就不行了。这时系统提供了将注入的 Option 进行命名的方式。

13.1.2 为 Option 命名的方式

需要在 Startup 中注册时对其命名，添加如下两条注册代码：

```
services.Configure<Theme>("ThemeBlue", Configuration.GetSection("Themes:0"));
services.Configure<Theme>("ThemeRed" , Configuration.GetSection("Themes:1"));
```

修改 ValueController 代码，添加 IOptionsMonitor<Theme>和 IOptionsSnapshot<Theme>两种新的注入方式，如下：

```
        private IOptions<Theme> _options = null;
        private IOptionsMonitor<Theme> _optionsMonitor = null;
        private IOptionsSnapshot<Theme> _optionsSnapshot = null;
        public ValuesController(IOptions<Theme> options, IOptionsMonitor<Theme>
optionsMonitor, IOptionsSnapshot<Theme> optionsSnapshot)
```

```
    {
        _options = options;
        _optionsMonitor = optionsMonitor;
        _optionsSnapshot = optionsSnapshot;
    }

    public ContentResult GetOptions()
    {
        return new ContentResult() { Content = $"options:{_options.Value.Name},"
+
            $"optionsSnapshot:{ _optionsSnapshot.Get("ThemeBlue").Name },"  +
            $"optionsMonitor:{_optionsMonitor.Get("ThemeRed").Name}" };
    }
```

请求 Action，获取的结果为：

```
options:Blue,optionsSnapshot:Red,optionsMonitor:Gray
```

新增的两种注入方式通过 Options 的名称获取了对应的 Options。为什么是两种呢？它们有什么区别？在配置注册时，有一个 reloadOnChange 参数，如果它被设置为 true，当对应的数据源发生改变时，会进行重新加载。而 Options 怎么能少了这样的特性呢？

13.1.3　Option 的自动更新与生命周期

为了验证这 3 种 Options 读取方式的特性，修改 Theme，添加一个 Guid 字段，并在构造方法中对其赋值，代码如下：

```
public class Theme
{
    public Theme()
    {
        Guid = Guid.NewGuid();
    }
    public Guid Guid { get; set; }
    public string Name { get; set; }
    public string Color { get; set; }
}
```

修改上例中的名为 GetOptions 的 Action 的代码，如下：

```
public ContentResult GetOptions()
{
    return new ContentResult()
    {
        Content = $"options:{_options.Value.Name}|{_options.Value.Guid}," +
$"optionsSnapshot:{ _optionsSnapshot.Get("ThemeBlue").Name }|{_optionsSnapshot.Get("The
meBlue").Guid}," +
```

```
$"optionsMonitor:{_optionsMonitor.Get("ThemeRed").Name}|{_optionsMonitor.Get
("ThemeRed").Guid}"
    };
}
```

请求 Action，返回结果如下：

```
options:Blue|ad328f15-254f-4505-a79f-4f27db4a393e,optionsSnapshot:Red|dba5f550-29ca-
4779-9a02-781dd17f595a,optionsMonitor:Gray|a799fa41-9444-45dd-b51b-fcd15049f98f
```

刷新页面，返回结果为：

```
options:Blue|ad328f15-254f-4505-a79f-4f27db4a393e,optionsSnapshot:Red|a2350cb3-c156-
4f71-bb2d-25890fe08bec,optionsMonitor:Gray|a799fa41-9444-45dd-b51b-fcd15049f98f
```

可见通过 IOptions 和 IOptionsMonitor 两种方式获取的 Name 值和 Guid 值均未改变；而通过 IOptionsSnapshot 方式获取的 Name 值未改变，Guid 值发生了改变，并且每次刷新页面均会改变。这类似上文讲依赖注入时做测试的例子，现在推测 Guid 值未改变的 IOptions 和 IOptionsMonitor 两种方式是采用了 Singleton 模式，而 Guid 值发生改变的 IOptionsSnapshot 方式是采用了 Scoped 或 Transient 模式。如果在这个 Action 中多次采用 IOptionsSnapshot 读取 _optionsSnapshot.Get("ThemeBlue").Guid 的值，会发现同一次请求的值是相同的，不同请求之间的值是不同的，也就是 IOptionsSnapshot 方式采用了 Scoped 模式（此验证示例比较简单，请读者自行修改代码验证）。

在这样的情况下，修改 3 种获取方式对应的配置项的 Name 值，例如分别修改为"Blue1""Red1"和"Gray1"，再多次刷新页面查看返回值，会发现如下情况。

- **IOptions 方式：** Name 和 Guid 的值始终未变，Name 值仍为 Blue。
- **IOptionsSnapshot 方式：** Name 值变为 Red1，Guid 值单次请求内相同，每次刷新之间不同。
- **IOptionsMonitor 方式：** 只有修改配置值后第一次刷新时，Name 值变为了 Gray1，Guid 值未改变。之后多次刷新，这两个值均未改变。

总结： IOptions 和 IOptionsMonitor 两种方式采用了 Singleton 模式，两者的区别在于 IOptionsMonitor 会监听对应数据源的变化。如果发生了变化，则更新实例的配置值，但不会重新提供新的实例。IOptionsSnapshot 方式采用了 Scoped 模式，每次请求采用同一个实例，在下一次请求时获取一个新的实例，所以如果数据源发生了改变，会读取新的值。读者先大概了解这些情况，在下文解析 IOptions 的内部处理机制时会详细解释其原因。

13.1.4 数据更新提醒

IOptionsMonitor 方式还提供了一个 OnChange 方法，当数据源发生改变时会触发它，所以如果想在这时做点什么，可以利用这个方法实现。示例代码如下：

```
_optionsMonitor.OnChange((theme,name)=> { Console.WriteLine(theme.Name +"-"+ name); });
```

13.1.5 其他配置方式

1. 不采用Configuration配置作为数据源的方式

上面的例子都采用了读取配置的方式，实际上Options模式和上一章的Configuration配置方式是分开的，读取配置只是Options模式的一种实现方式。例如，可以不使用Configuration中的数据，直接通过如下代码注册：

```
services.Configure<Theme>("ThemeBlack", theme => {
    theme.Color = "#000000";
    theme.Name = "Black";
});
```

2. ConfigureAll方法

系统提供了ConfigureAll方法，可以将所有对应的实例统一设置。例如以下代码：

```
services.ConfigureAll<Theme>(theme => {
    theme.Color = "#000000";
    theme.Name = "Black2";
});
```

此时无论通过什么名称去获取Theme的实例，包括不存在对应设置的名称，获取的都是本次通过ConfigureAll设置的值。

3. PostConfigure和PostConfigureAll方法

这两个方法和Configure、ConfigureAll方法类似，只是它们会在Configure、ConfigureAll之后执行。

4. 多个Configure、ConfigureAll、PostConfigure和PostConfigureAll的执行顺序

可以这样理解——每个Configure都是去修改一个名为其设置的名称的变量，以如下代码为例：

```
services.Configure<Theme>("ThemeBlack", theme => {
    theme.Color = "#000000";
    theme.Name = "Black";
});
```

这段代码的设置的作用是修改（注意是修改而不是替换）一个名为ThemeBlack的Theme的变量，如果该变量不存在，则创建一个Theme实例并赋值。这样就生成了一些变量名为“空字符串”ThemeBlue ThemeBlack的变量（只是举例，忽略空字符串作为变量名不合法的顾虑）。

依次按照代码的顺序执行，这时如果后面的代码中出现同名的Configure，则修改对应名称的变量的值。如果是ConfigureAll方法，则修改所有Theme。

而 PostConfigure 和 PostConfigureAll 则在 Configure 和 ConfigureAll 之后执行，即使 Configure 的代码写在了 PostConfigure 之后也是一样。至于为什么会是这样的规则，下一节会详细介绍。

13.2 内部处理机制解析

13.2.1　系统启动阶段的依赖注入

上一节的例子中涉及了 3 个接口——IOptions、IOptionsSnapshot 和 IoptionsMonitor。既然 Options 模式是通过这 3 个接口的泛型方式注入提供服务的，那么在这之前，系统就需要将它们对应的实现注入依赖注入容器中。这发生在系统启动阶段创建 IHost 时，HostBuilder 的 Build 方法中调用了一个 AddOptions 方法，该方法定义在 OptionsServiceCollectionExtensions 中，代码如下：

```
public static class OptionsServiceCollectionExtensions
    {
        public static IServiceCollection AddOptions(this IServiceCollection services)
        {
            if (services == null)
            {
                throw new ArgumentNullException(nameof(services));
            }

            services.TryAdd(ServiceDescriptor.Singleton(typeof(IOptions<>), typeof
(OptionsManager<>)));
            services.TryAdd(ServiceDescriptor.Scoped(typeof(IOptionsSnapshot<>),
typeof(OptionsManager<>)));
            services.TryAdd(ServiceDescriptor.Singleton(typeof(IOptionsMonitor<>),
typeof(OptionsMonitor<>)));
            services.TryAdd(ServiceDescriptor.Transient(typeof(IOptionsFactory<>),
typeof(OptionsFactory<>)));
            services.TryAdd(ServiceDescriptor.Singleton(typeof(IOptionsMonitorCache<>),
typeof(OptionsCache<>)));
            return services;
        }

        public static IServiceCollection Configure<TOptions>(this IServiceCollection
services, Action<TOptions> configureOptions) where TOptions : class
            => services.Configure(Options.Options.DefaultName, configureOptions);

        public static IServiceCollection Configure<TOptions>(this IServiceCollection
services, string name, Action<TOptions> configureOptions)
            where TOptions : class
        {
            //省略非空验证代码
```

```
            services.AddOptions();
            services.AddSingleton<IConfigureOptions<TOptions>>(new ConfigureNamedOptions
<TOptions>(name, configureOptions));
            return services;
        }

        public static IServiceCollection ConfigureAll<TOptions>(this IServiceCollection
services, Action<TOptions> configureOptions) where TOptions : class
            => services.Configure(name: null, configureOptions: configureOptions);
//省略部分代码
}
```

可见 AddOptions 方法的作用是进行服务注入，IOptions、IOptionsSnapshot 对应的实现是 OptionsManager，只是分别采用了 Singleton 和 Scoped 两种模式。IOptionsMonitor 对应的实现是 OptionsMonitor，同样为 Singleton 模式，这也验证了上一节例子中的猜想。除了上面提到的 3 个接口，还有 IOptionsFactory 和 IOptionsMonitorCache 两个接口也是 Options 模式中非常重要的组成部分，接下来的内容中会用到。

另两个 Configure 方法是上一节例子中用到的将具体的 Theme 注册到 Options 中的方法。二者的区别是为配置的 Option 命名。第一个 Configure 方法就是未命名的方法，通过上面的代码可知，它实际上传入了一个默认的 Options.Options.DefaultName 作为名称，这个默认值是一个空字符串，也就是说，未命名的 Options 相当于被命名为空字符串，最终都是按照已命名的方式，也就是第二个 Configure 方法进行处理。还有一个 ConfigureAll 方法，传入了一个 Null 作为 Options 的名称，也是交给第二个 Configure 方法处理。

在第二个 Configure 方法中调用一次 AddOptions 方法，然后才将具体的 Theme 进行注入。AddOptions 方法中都是调用 TryAdd 方法来进行注入，已被注入的不会被再次注入。接下来注册了一个 IConfigureOptions<TOptions>接口，对应的实现是 ConfigureNamedOptions<TOptions>(name, configureOptions)，代码如下：

```
public class ConfigureNamedOptions<TOptions> : IConfigureNamedOptions<TOptions> where
 TOptions : class
{
    public ConfigureNamedOptions(string name, Action<TOptions> action)
    {
        Name = name;
        Action = action;
}

    public string Name { get; }
    public Action<TOptions> Action { get; }

    public virtual void Configure(string name, TOptions options)
```

```
    {
        if (options == null)
        {
            throw new ArgumentNullException(nameof(options));
        }

        if (Name == null || name == Name)
        {
            Action?.Invoke(options);
        }
    }

    public void Configure(TOptions options) => Configure(Options.DefaultName, options);
}
```

该接口在构造方法中存储了配置的名称（Name）和创建方法（Action），两个 Configure 方法用于在获取 Options 的值时执行对应的 Action 来创建实例（例如示例中的 Theme），并且此时不会被执行。所以在此会出现 3 种类型的 ConfigureNamedOptions，分别是 Name 值为具体值的、Name 值为空字符串的和 Name 值为 Null 的，这分别对应了 13.1 节例子中的为 Options 命名的 Configure 方法、不为 Options 命名的 Configure 方法，以及 ConfigureAll 方法。

此处用到的 OptionsServiceCollectionExtensions 和 ConfigureNamedOptions 对应的是通过代码直接注册 Options 的方式，例如 13.1 节例子中的以下方式：

```
services.Configure<Theme>("ThemeBlack", theme => { new Theme { Color = "#000000",
Name = "Black" }; });
```

如果是以 Configuration 作为数据源的方式，则代码如下：

```
services.Configure<Theme>("ThemeBlue", Configuration.GetSection("Themes:0"));
```

用到了 OptionsServiceCollectionExtensions 和 ConfigureNamedOptions 这两个类的子类，分别为 OptionsConfigurationServiceCollectionExtensions 和 NamedConfigureFromConfigurationOptions 两个类，通过名字可以知道，它们是专门使用 Configuration 作为数据源的。其代码只是多了一条关于 IOptionsChangeTokenSource 的依赖注入，作用是将 Configuration 的关于数据源变化的监听和 Options 的关联起来，当数据源发生改变时可以及时更新 Options 中的值，主要的 Configure 方法代码如下：

```
public static IServiceCollection Configure<TOptions>(this IServiceCollection services,
string name, IConfiguration config, Action<BinderOptions> configureBinder)
    where TOptions : class
{
    //省略验证代码

    services.AddOptions();
```

```
    services.AddSingleton<IOptionsChangeTokenSource<TOptions>>(new Configuration
ChangeTokenSource<TOptions>(name, config));
    return services.AddSingleton<IConfigureOptions<TOptions>>(new NamedConfigureFrom
ConfigurationOptions<TOptions>(name, config, configureBinder));
}
```

同样还有 PostConfigure 和 PostConfigureAll 方法，它们与 Configure、ConfigureAll 方法类似，但注入的类型为 IPostConfigureOptions<TOptions>。

13.2.2　Options 值的获取

Options 值的获取也就是从依赖注入容器中获取相应实现的过程。通过依赖注入阶段可知，IOptions 和 IOptionsSnapshot 对应的实现是 OptionsManager，现在以 OptionsManager 为例看依赖注入后的服务提供过程。OptionsManager 的代码如下：

```
public class OptionsManager<TOptions> : IOptions<TOptions>, IOptionsSnapshot<TOptions>
where TOptions : class, new()
{
    private readonly IOptionsFactory<TOptions> _factory;
    private readonly OptionsCache<TOptions> _cache = new OptionsCache<TOptions>();

    public OptionsManager(IOptionsFactory<TOptions> factory)
    {
        _factory = factory;
    }

    public TOptions Value
    {
        get
        {
            return Get(Options.DefaultName);
        }
    }

    public virtual TOptions Get(string name)
    {
        name = name ?? Options.DefaultName;
        return _cache.GetOrAdd(name, () => _factory.Create(name));
    }
}
```

其中有 IOptionsFactory<TOptions>和 OptionsCache<TOptions>两个重要的成员。如果直接获取 Value 值，实际上是调用另一个 Get(string name)方法，传入了空字符串作为 Name 值，所以最终值的获取还是在缓存中读取。这里的代码是_cache.GetOrAdd(name, () =>_factory.Create(name))，即如果缓存中存在对应的值，则返回；如果不存在，则由_factory 去创建。OptionsFactory<TOptions>

的代码如下：

```
public class OptionsFactory<TOptions> : IOptionsFactory<TOptions> where TOptions : class,
 new()
{
    private readonly IEnumerable<IConfigureOptions<TOptions>> _setups;
    private readonly IEnumerable<IPostConfigureOptions<TOptions>> _postConfigures;
    private readonly IEnumerable<IValidateOptions<TOptions>> _validations;

    public OptionsFactory(IEnumerable<IConfigureOptions<TOptions>> setups,
Ienumerable<IPostConfigureOptions<TOptions>> postConfigures) :
this(setups, postConfigures, validations:null)
    { }

    public OptionsFactory(IEnumerable<IConfigureOptions<TOptions>> setups,
Ienumerable<IPostConfigureOptions<TOptions>> postConfigures,
IEnumerable<IValidateOptions<TOptions>> validations)
    {
        _setups = setups;
        _postConfigures = postConfigures;
        _validations = validations;
}

    public TOptions Create(string name)
    {
        var options = new TOptions();
        foreach (var setup in _setups)
        {
            if (setup is IConfigureNamedOptions<TOptions> namedSetup)
            {
                namedSetup.Configure(name, options);
            }
            else if (name == Options.DefaultName)
            {
                setup.Configure(options);
            }
        }
        foreach (var post in _postConfigures)
        {
            post.PostConfigure(name, options);
        }

        if (_validations != null)
        {
            var failures = new List<string>();
            foreach (var validate in _validations)
```

```
                {
                    var result = validate.Validate(name, options);
                    if (result.Failed)
                    {
                        failures.AddRange(result.Failures);
                    }
                }
                if (failures.Count > 0)
                {
                    throw new OptionsValidationException(name, typeof(TOptions), failures);
                }
            }

            return options;
        }
    }
```

其中，TOptions Create(string name)方法会遍历_setups 集合，集合类型为 Ienumerable<IConfigureOptions<TOptions>>。在讲 Options 模式的依赖注入时可知，每一个 Configure、ConfigureAll 方法实际上是向依赖注入容器中注册一个 IConfigureOptions<TOptions>，只是名称可能不同。而 PostConfigure 和 PostConfigureAll 方法注册的是 IPostConfigureOptions<TOptions>类型，对应_postConfigures 集合。

遍历_setups 集合，调用 IConfigureOptions<TOptions>的 Configure 方法，主要代码如下：

```
if (Name == null || name == Name)
{
    Action?.Invoke(options);
}
```

如果 Name 值为 Null，即对应 ConfigureAll 方法，执行该 Action。如果 Name 值和需要读取的值相同，则执行该 Action。

_setups 集合遍历之后，以同样的机制遍历_postConfigures 集合。这就是 13.1 节关于 Configure、ConfigureAll、PostConfigure 和 PostConfigureAll 的执行顺序的验证。

最终返回对应的实例并写入缓存，这就是 IOptions 和 IOptionsSnapshot 两种模式的处理机制。接下来看 IOptionsMonitor 模式，对应的实现是 OptionsMonitor，代码如下：

```
public class OptionsMonitor<TOptions> : IOptionsMonitor<TOptions> where TOptions :
class, new()
{
    private readonly IOptionsMonitorCache<TOptions> _cache;
    private readonly IOptionsFactory<TOptions> _factory;
    private readonly IEnumerable<IOptionsChangeTokenSource<TOptions>> _sources;
    internal event Action<TOptions, string> _onChange;
```

```
    public OptionsMonitor(IOptionsFactory<TOptions> factory,
IEnumerable<Ioptions ChangeTokenSource<TOptions>> sources,
IOptionsMonitorCache<TOptions> cache)
    {
        _factory = factory;
        _sources = sources;
        _cache = cache;

        foreach (var source in _sources)
        {
                var registration = ChangeToken.OnChange(
                      () => source.GetChangeToken(),
                      (name) => InvokeChanged(name),
                      source.Name);

                _registrations.Add(registration);
        }
    }

    private void InvokeChanged(string name)
    {
        name = name ?? Options.DefaultName;
        _cache.TryRemove(name);
        var options = Get(name);
        if (_onChange != null)
        {
            _onChange.Invoke(options, name);
        }
    }

    public TOptions CurrentValue
    {
        get => Get(Options.DefaultName);
    }

    public virtual TOptions Get(string name)
    {
        name = name ?? Options.DefaultName;
        return _cache.GetOrAdd(name, () => _factory.Create(name));
    }

    public IDisposable OnChange(Action<TOptions, string> listener)
    {
        var disposable = new ChangeTrackerDisposable(this, listener);
        _onChange += disposable.OnChange;
        return disposable;
```

```
    }

    internal class ChangeTrackerDisposable : IDisposable
    {
        private readonly Action<TOptions, string> _listener;
        private readonly OptionsMonitor<TOptions> _monitor;

        public ChangeTrackerDisposable(OptionsMonitor<TOptions> monitor, Action
<TOptions, string> listener)
        {
            _listener = listener;
            _monitor = monitor;
        }

        public void OnChange(TOptions options, string name) => _listener.Invoke
(options, name);

        public void Dispose() => _monitor._onChange -= OnChange;
    }
}
```

上述代码的大部分功能和 OptionsManager 类似，只是采用了 Singleton 模式，所以它采用监听数据源改变并更新的模式。当通过 Configuration 作为数据源注册 Option 时，多了一条 IOptionsChangeTokenSource 的依赖注入；当数据源发生改变时，更新数据并触发 OnChange(Action<TOptions, string> listener)，在 13.1 节的数据更新提醒中有相关的例子。

第 14 章　请求处理管道

本章讲解 ASP.NET Core 的重要组成部分之一——请求处理管道，包括管道的配置、构建，以及请求处理流程等方面。

14.1 概述

在第 7 章讲到，请求需要经过 Server 监听获取，然后被处理成 httpContext，最终被 Application 处理生成 Response 过程，而这个处理过程主要由请求处理管道来完成（见图 7-2）。请求处理管道由多个中间件组成，如图 14-1 所示。

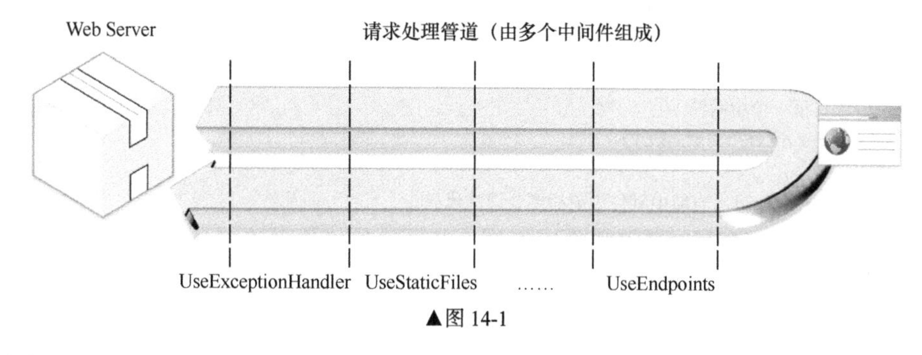

▲图 14-1

用户如果要获取 Web 页面或其他数据，可以发送一个 HTTP 请求到服务器，过程中会经过一个个中间件，获取结果之后，反向通过中间件直至返回给用户。这就是 ASP.NET Core 的一次请求的接收与处理过程。

这一个个中间件组成了 ASP.NET Core 的请求处理管道，整条管道就像高速公路，使请求从公路的一侧行驶到达目的地，又从公路的另一侧返回出发地。在请求发起前，这条管道就已经构建完成，管道的构建发生在应用启动阶段。

本章分为“请求处理管道的构建”和“请求到达管道并被处理”两个阶段来讲解。

14.2 请求在管道中的处理流程

为了方便理解，下面通过一个简单的例子讲解中间件，再通过构建一个简易的请求处理管道了解请求处理的过程。

14.2.1　简单的中间件例子

在图 14-1 中，每一个由虚线分割的一段“管道”代表一个中间件（Middleware），请求会经过每个中间件两次。中间件的类型是 Func<RequestDelegate, RequestDelegate>，这个类型可能不是很好理解，下面看一个简单的中间件例子：

```
public class OneMiddleware
{
    private readonly RequestDelegate _next;
    public OneMiddleware(RequestDelegate next)
    {
        _next = next;
    }
    public async Task InvokeAsync(HttpContext context)
    {
        #region //进入当前中间件后需要处理的代码
        Console.WriteLine("OneMiddleware In");
        #endregion

        //去往下一个中间件
        await _next(context);

        #region //退出当前中间件之前需要处理的代码
        Console.WriteLine("OneMiddleware Out");
        #endregion

    }
}
```

其中，第一个#region 就是进入第一个中间件时的操作语句，中间的_next(context)则是出了当前中间件进入下一个中间件，最后一个 region 则是从进入的下一个中间件回到先前中间件时的操作。以高速公路为例，一个中间件相当于一个地区，去和回的行程一共会经过这个地区两次，所以每个中间件可以有两次处理请求的机会。

14.2.2　请求是如何经过各个中间件的

创建中间件之后，需要将其添加到请求处理管道才能生效，此过程可由 Startup 中的 Configure 方法实现，这个方法可供我们按照一定顺序将一系列中间件构建成请求处理管道，默认情况下的代码如下：

```
public void Configure(IApplicationBuilder app, IWebHostEnvironment env)
{
    if (env.IsDevelopment())
    {
        app.UseDeveloperExceptionPage();
    }
    else
    {
        app.UseExceptionHandler("/Home/Error");
}

    app.UseStaticFiles();

    app.UseRouting();

    app.UseAuthorization();

    app.UseEndpoints(endpoints =>
    {
        endpoints.MapControllerRoute(
            name: "default",
            pattern: "{controller=Home}/{action=Index}/{id?}");
    });
}
```

该方法中的每个Use*XXX*中包含了一个或多个中间件的设置，此处为了方便理解和管理而放在了一起。UseStaticFiles的代码如下：

```
public static IApplicationBuilder UseStaticFiles(this IApplicationBuilder app)
{
    if (app == null)
    {
        throw new ArgumentNullException(nameof(app));
    }

    return app.UseMiddleware<StaticFileMiddleware>();
}
```

Startup的Configure方法依次设置了Exception、静态文件、路由、授权等相关的中间件。多个中间件按照一定的顺序组合在一起，形成请求处理管道。

仿照OneMiddleware再次新建两个中间件TwoMiddleware和ThreeMiddleware，并将这3个中间件添加到管道中，代码如下：

```
public void Configure(IApplicationBuilder app, IWebHostEnvironment env)
```

```
{
    app.UseMiddleware<OneMiddleware>();
    app.UseMiddleware<TwoMiddleware>();
    app.UseMiddleware<ThreeMiddleware>();

    if (env.IsDevelopment())
    //省略了之后的代码
}
```

将它们放在最前面，所以请求会最先进入这 3 个中间件。运行项目后会看到输出的日志如下所示：

```
OneMiddleware In
TwoMiddleware In
ThreeMiddleware In
info: Microsoft.AspNetCore.Routing.EndpointMiddleware[0]
      Executing endpoint 'HelloCore.Controllers.HomeController.Index (HelloCore)'
//省略其他输出代码
ThreeMiddleware Out
TwoMiddleware Out
OneMiddleware Out
```

这样输出的日志验证了请求经过中间件的顺序。

14.3 管道的构建

了解了多个中间件的嵌套方式后，读者可能会有疑问：Startup 中的 Configure 方法是如何构建请求处理管道的呢？为什么中间件的_next(context)方法会执行进入下一个中间件呢？

这个嵌套过程发生在 Host 启动时，具体是在 GenericWebHostService 的 StartAsync 方法中，在第 8 章涉及此部分内容时未做详细介绍。简化后的这段代码如下：

```
internal class GenericWebHostService : IHostedService
{
    public async Task StartAsync(CancellationToken cancellationToken)
    {
        RequestDelegate application = null;
        Action<IApplicationBuilder> configure = Options.ConfigureApplication;

        var builder = ApplicationBuilderFactory.CreateBuilder(Server.Features);

        foreach (var filter in StartupFilters.Reverse())
        {
            configure = filter.Configure(configure);
        }
```

```
        configure(builder);

        application = builder.Build();
        //省略部分代码
    }
}
```

首先 Action<IApplicationBuilder> 类型的 configure 被赋值为 Options.ConfigureApplication，在第 8 章讲 Startup 文件的处理时提到，Options.ConfigureApplication 就是 Startup 的 Configure 方法。

接下来创建一个 ApplicationBuilder 类的 builder，暂时跳过。foreach 遍历 StartupFilters，它是一个 IStartupFilter 集合，IStartupFilter 接口代码如下：

```
public interface IStartupFilter
{
    Action<IApplicationBuilder> Configure(Action<IApplicationBuilder> next);
}
```

IStartupFilter 接口只有一个 Configure 方法，这个方法名会让我们想起 Startup 的 Configure 方法，其实这两个方法的作用是相似的。默认情况下，StartupFilters 包含了两个子类，即 HostFilteringStartupFilter 和 MiddlewareFilterBuilderStartupFilter。以前者为例，其代码如下：

```
internal class HostFilteringStartupFilter : IStartupFilter
{
    public Action<IApplicationBuilder> Configure(Action<IApplicationBuilder> next)
    {
        return app =>
        {
            app.UseHostFiltering();
            next(app);
        };
    }
}
```

可以看出 IStartupFilter 的 Configure 方法的作用同样是添加中间件，但不是立即添加，而是将添加中间件的方法通过调用 next(app)和传入的参数 next 拼接在一起。可以说，IStartupFilter 的存在就是为了给我们多一种添加中间件的途径。

通过 foreach 遍历，将 StartupFilters 中的一个或多个 IStartupFilter 子类的 Configure 方法与 Configure 变量（Startup 的 Configure 方法）拼接在一起，然后调用 Configure(builder)方法。

接下来的工作是由 builder 完成的，其类型是 ApplicationBuilder，ApplicationBuilder 简化后的代码如下：

```
public class ApplicationBuilder : IApplicationBuilder
{
```

```
    private readonly IList<Func<RequestDelegate, RequestDelegate>> _components = new
List<Func<RequestDelegate, RequestDelegate>>();

    public IApplicationBuilder Use(Func<RequestDelegate, RequestDelegate> middleware)
    {
        _components.Add(middleware);
        return this;
    }

    public RequestDelegate Build()
    {
        RequestDelegate app = context =>
        {
            var endpoint = context.GetEndpoint();
            var endpointRequestDelegate = endpoint?.RequestDelegate;
            if (endpointRequestDelegate != null)
            {
                var message = "错误提示信息略";
                throw new InvalidOperationException(message);
            }

            context.Response.StatusCode = 404;
            return Task.CompletedTask;
        };

        foreach (var component in _components.Reverse())
        {
            app = component(app);
        }

        return app;
    }
}
```

ApplicationBuilder 中有集合 IList<Func<RequestDelegate, RequestDelegate>>_components 和一个用于向这个集合中添加内容的 Use(Func<RequestDelegate, RequestDelegate> middleware)方法，通过类型可以看出它们是用来添加和存储中间件的。

Configure(builder)是调用 StartupFilters 和 Startup 拼接成的 Configure 方法，调用其中定义的多个 Use*XXX* 将一个个中间件 middleware 按照顺序写入 builder 的_components 集合。

接下来调用 builder 的 Build 方法，首先定义一个 context.Response.StatusCode = 404 的 RequestDelegate，然后将_components 集合中间件的配置顺序颠倒一下，并遍历其中的 Middleware，将一个个的 Middleware 与新创建的 StatueCode=404 的 RequestDelegate 连接在一起，组成一个新的 RequestDelegate（即 Application）返回。最终返回的 RequestDelegate 的 Application 就是对 HttpContext 处理的管道了，一个“中规中矩”的管道就是这样构建并运行

的。图 14-2 描述了这一过程。

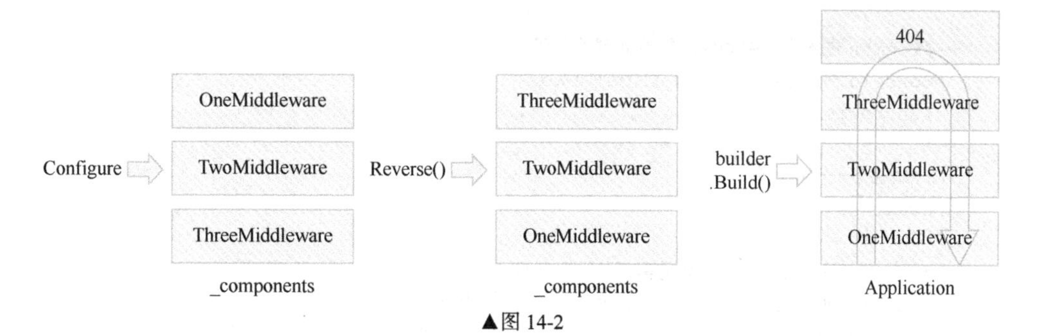

▲图 14-2

从图 14-2 中可以看到，各个中间件在 Startup 文件中的配置顺序与最终构成的管道中的顺序的关系。为什么说这是一个“中规中矩”的管道呢？因为还可能是“分叉”的请求处理管道，下面我们通过创建几个中间件来体验一下。

14.4 中间件的其他定义方式

在 14.2 节的例子中，我们了解到 StaticFiles 相关中间件是通过扩展方法 UseStaticFiles 注册的，好处是名字易于理解，也可以使 Startup 的 Configure 方法中的代码更简洁易懂。一个功能由多个中间件组成，那么可以将这几个中间件的注册写在同一个扩展方法中。我们也试着将 OneMiddleware、TwoMiddleware 等自定义中间件通过扩展方法注册。定义一个名为 UseSimple 的扩展方法，代码如下：

```
public static class SimpleMiddlewareExtensions
{
    public static IApplicationBuilder UseSimple(this IApplicationBuilder builder)
    {
        Console.WriteLine("Use Simple");
        return builder.UseMiddleware<OneMiddleware>()
                      .UseMiddleware<TwoMiddleware>()
                      .UseMiddleware<ThreeMiddleware>();
    }
}
```

这样在 Startup 的 Configure 方法中也可以写 UseSimple 方法，并将这个中间件作为管道的一部分。

通过上面的例子，仿照系统默认的中间件完成一个简单的中间件的编写，这里也可以用简要的方法，直接在 Startup 的 Configure 方法中写为：

```
public void Configure(IApplicationBuilder app, IWebHostEnvironment env)
{
```

```
    app.UseSimple();
    app.Use(async (context, next) =>
    {
        #region 进入当前中间件后需要处理的代码
        Console.WriteLine("FourMiddleware In");
        #endregion

        //去往下一个中间件
        await next.Invoke();

        #region 退出当前中间件之前需要处理的代码
        Console.WriteLine("FourMiddleware Out");
        #endregion
    });
}
```

上述代码同样可以实现中间件的工作，但还是建议读者采用扩展方法实现。

14.5 Use、Run 和 Map

14.5.1 Use 和 Run

在 14.3 节讲管道的构建时，系统默认添加一个返回代码为 404 的中间件，在前面的中间件都未对请求做返回结果处理的情况下，就会到达这个中间件。把 Configure 方法中除自定义的 4 个中间件以外的中间件全部注释掉，再次运行，可以看到如图 14-3 所示的输出日志，返回了 404。

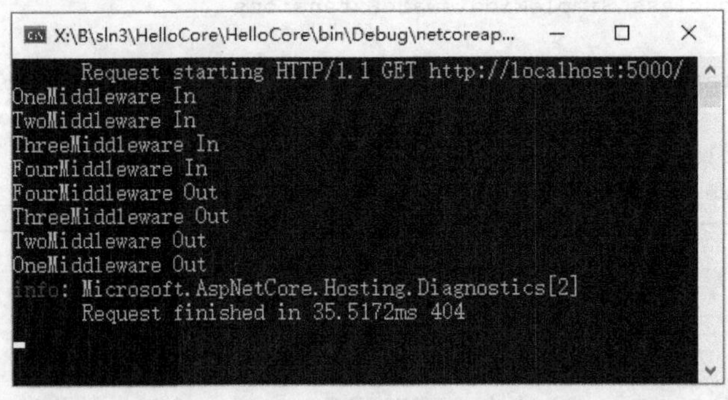

▲图 14-3

可以看到，MVC 处理的部分没有了，因为该中间件已被注释掉，而从最后一条可以看到系统返回了 404。

既然 MVC 可以正常处理请求而没有进入状态码为 404 的 RequestDelegate，是不是不调用下一个中间件了？试着把 FourMiddleware 改为：

```
app.Use(async (context, next) =>
```

```
{
    #region 进入当前中间件后需要处理的代码
    Console.WriteLine("FourMiddleware In");
    #endregion

    //去往下一个中间件
    //await next.Invoke();
    await context.Response.WriteAsync("Danger!");

    #region 退出当前中间件之前需要处理的代码
    Console.WriteLine("FourMiddleware Out");
    #endregion
});
```

再次运行，发现输出日志和上文的日志没有太大差别，只是最后的 404 变为了 200。查看网页已经按照预期输出了“Danger!”，达到了我们想要的效果。但一般情况下我们不这样写，而是以 ASP.NET Core 提供的 Use、Run 和 Map 三种方法来配置管道。一般用 Run 方法实现，Run 方法主要用来作为管道的末尾，例如上面的代码可以改写为：

```
app.Run(async (context) =>
{
    await context.Response.WriteAsync("Danger!");
});
```

Run 方法作为管道末尾，也就省略了 next 参数。虽然用 Use 方法也可以实现，但还是建议读者用 Run 方法。

14.5.2 Map

Map 方法代码如下：

```
    static IApplicationBuilder Map(this IApplicationBuilder app, PathString pathMatch,
Action<IApplicationBuilder> configuration);
```

pathMatch 用于匹配请求的 path，例如“/Home”，且必须以“/”开头。判断 path 是否以 pathMatch 开头，若是，则进入 Action<IApplicationBuilder> configuration，这个参数和 Startup 的 Configure 方法很像，就像进入了我们配置的另一个管道，它是一个分支，如图 14-4 所示。

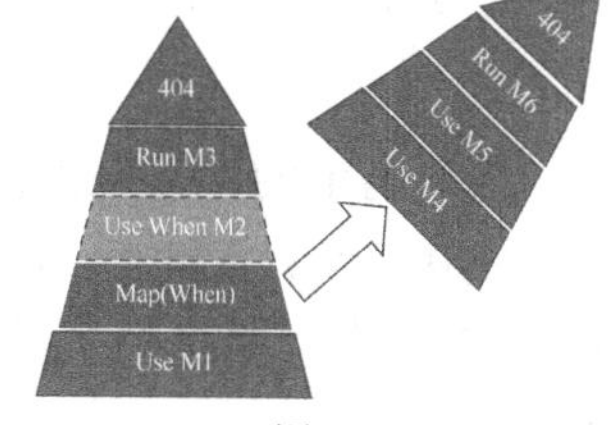

▲图 14-4

举个例子：

```
public void Configure(IApplicationBuilder app, IWebHostEnvironment env)
{
    app.UseSimple();
```

```
    app.Map("/Manager", builder =>
    {
        builder.Use(async (context, next) =>
        {
            Console.WriteLine("FiveMiddleware In");
            await next.Invoke();
            Console.WriteLine("FiveMiddleware out");
        });

        builder.Run(async (context) =>
        {
            await context.Response.WriteAsync("Manager.");
        });
    });

    app.Run(async (context) =>
    {
        await context.Response.WriteAsync("Danger!");
    });
}
```

当我们请求 Manager/Index 这样的地址时（有点像 Area），会进入这个 Map 创建的新分支，结果是页面显示“Manager.”，不会再进入后面的 app.Run()中间件。若不是以“/Manager”开头，则继续进入 app.Run()中间件，结果返回“Danger!”。虽然感觉 Map 使管道配置变灵活了，但只能匹配 path 开头的方法太有局限性了。下面来看 MapWhen。

14.5.3 MapWhen

MapWhen 方法是一个灵活版的 Map，它将原来的 pathMatch 替换为一个 Func<HttpContext, bool> predicate，返回一个 bool 值。将方法修改为：

```
app.MapWhen(context=> {return context.Request.Query.ContainsKey("ss");}, builder =>
{
  //...TODO...
}
```

根据请求的参数是否包含“ss”判断是否进入这个分支。

由图 14-4 可知，一旦进入分支，是无法回到原分支的。如果只是想在某种情况下进入某些中间件，但执行完后还想继续后续的中间件怎么办呢？对比 MapWhen，Use 也有 UseWhen。

14.5.4 UseWhen

UseWhen 和 MapWhen 一样，当满足条件时进入一个分支，在这个分支完成之后再继续后续的中间件，前提是这个分支中没有使用 Run 方法结尾。代码如下：

```
app.UseWhen(context=> {return context.Request.Query.ContainsKey("ss");}, builder =>
{
    //...TODO...
}
```

14.6 IStartupFilter

我们只能指定一个 Startup 类作为启动类，那么还能在其他的地方定义管道吗？构建管道时会遍历 StartupFilters 集合中的 IStartupFilter，并逐一执行 Configure 方法，调用其中定义的多个 Use*XXX* 方法将中间件写入_components 集合。

我们也可以自定义一个 StartupFilter，实现 IStartupFilter 的 Configure 方法，用法和 Startup 的 Configure 类似，最后要记得调用 next(app)。代码如下：

```
    public class TestStartupFilter : IStartupFilter
    {
        public Action<IApplicationBuilder> Configure(Action<IApplicationBuilder> next)
        {
            return app =>
            {
                app.UseMiddleware<SixMiddleware>();
                app.UseMiddleware<SevenMiddleware>();
                next(app);
            };
        }
    }
```

用类似的方法再创建一个 TestStartupFilter2，并在 Startup 的 ConfigureServices 方法中注册：

```
services.AddSingleton<IStartupFilter,TestStartupFilter>();
services.AddSingleton<IStartupFilter,TestStartupFilter2>();
```

这样配置就生效了，下面剖析它的生效机制。回顾 GenericWebHostService 的 StartAsync 方法：

```
internal class GenericWebHostService : IHostedService
{
    public async Task StartAsync(CancellationToken cancellationToken)
    {
        foreach (var filter in StartupFilters.Reverse())
        {
            configure = filter.Configure(configure);
        }
        //省略部分代码
        }
}
```

这段代码其实和构建管道的流程非常相似，对比来说：首先，IstartupFilter 的 StartupFilters 集合类似_components；其次，Startup 的 Configure 类似管道构建时默认创建的状态码为 404 的 RequestDelegate。颠倒 StartupFilters 中的顺序，通过 foreach 遍历并且与 Startup 的 Configure 拼接在一起。

记得最后调用 next(app)，这和 next.Invoke() 也是类似的。整个过程和图 14-2 的翻转拼接过程非常相似，对比管道构建后中间件的执行顺序，可以想到各个 IStartupFilter 和 Startup 的 Configure 的执行顺序，没错，就是按照依赖注入的顺序：TestStartupFilter→TestStartupFilter2→Startup。

第 15 章　静态文件访问与授权

如何避免网站的图片被公开浏览、下载、盗链呢？本文主要通过解读 ASP.NET Core 对静态文件的处理方式的相关代码，来看什么是 wwwroot 文件目录。如何修改或新增一个静态文件夹？为什么新增的文件夹名字不会被当作 Controller 处理？怎么做访问授权？

15.1 静态文件夹

所谓静态文件，就是项目中 wwwroot 文件目录下的一些直接提供给访问者的文件，例如 CSS、图片、JS 文件等。例如，直接访问 http://localhost:5000/css/site.css 可以获取该 CSS 文件的内容，而不需要经过 Controller 处理。

这个 wwwroot 文件目录只是默认的静态文件目录，可以被修改，其默认设置是在 HostingEnvironmentExtensions 的 Initialize 方法中。在 GenericWebHostBuilder 被创建触发的构造方法中调用 GetWebHostBuilderContext 方法，进而将 HostingEnvironmentExtensions 的 Initialize 方法加入 HostBuilder 的_configureAppConfigActions 中执行。代码如下：

```
public static class HostingEnvironmentExtensions
    {
        public static void Initialize(this IHostingEnvironment hostingEnvironment,
string contentRootPath, WebHostOptions options)
        {
           //省略部分代码
            var webRoot = options.WebRoot;
            if (webRoot == null)
            {
                var wwwroot = Path.Combine(hostingEnvironment.ContentRootPath, "wwwroot");
                if (Directory.Exists(wwwroot))
                {
                    hostingEnvironment.WebRootPath = wwwroot;
                }
            }
            else
```

```
        {
            hostingEnvironment.WebRootPath =
Path.Combine(hostingEnvironment.ContentRootPath, webRoot);
        }
        //省略部分代码
    }
}
```

15.2 中间件的实现机制

在 Startup 中，有一个 UseStaticFiles 方法的调用，这里是将静态文件的处理中间件作为请求处理管道的一部分。这个中间件被放置得非常靠前，当一个请求进来之后，会先判断其是否为静态文件的请求。如果是，则在此做请求处理，这时请求会发生短路，不会进入后面的路由等中间件处理步骤。代码如下：

```
public void Configure(IApplicationBuilder app, IWebHostEnvironment env)
{

    if (env.IsDevelopment())
    {
        app.UseDeveloperExceptionPage();
    }
    else
    {
        app.UseExceptionHandler("/Home/Error");
    }
    app.UseStaticFiles();
    app.UseRouting();
    app.UseAuthorization();
    app.UseEndpoints(endpoints =>
    {
        endpoints.MapControllerRoute(
            name: "default",
            pattern: "{controller=Home}/{action=Index}/{id?}");
    });
}
```

静态文件的处理中间件为 StaticFileMiddleware，主要的处理方法是 Invoke，其代码如下：

```
public Task Invoke(HttpContext context)
{
    if (!ValidateNoEndpoint(context))
    {
        _logger.EndpointMatched();
    }
    else if (!ValidateMethod(context))
```

```
    {
        _logger.RequestMethodNotSupported(context.Request.Method);
    }
    else if (!ValidatePath(context, _matchUrl, out var subPath))
    {
        _logger.PathMismatch(subPath);
    }
    else if (!LookupContentType(_contentTypeProvider, _options, subPath, out var
contentType))
    {
        _logger.FileTypeNotSupported(subPath);
    }
    else
    {
        return TryServeStaticFile(context, contentType, subPath);
    }
    return _next(context);
}
```

HttpContext 进入此中间件后判断是否有匹配的终结点、是否是 GET 请求、请求的 URL 是否与设置的静态目录一致、请求头是否正确、文件类型是否支持等。若这些全部验证通过，会进一步判断文件是否存在、是否有修改等，最终将对应的静态文件作为请求结果返回。

15.3 新增静态文件目录

除了这个默认的 wwwroot 目录，还需要新增一个目录作为静态文件的目录，可以在 Startup 的 UseStaticFiles 下继续执行 use 方法，例如如下代码：

```
app.UseStaticFiles(new StaticFileOptions
{
    FileProvider = new PhysicalFileProvider(
        Path.Combine(Directory.GetCurrentDirectory(), "NewWwwroot")),
        RequestPath = "/NewFiles"
});
```

含义是指定应用程序目录中的一个名为“NewWwwroot”的文件夹，将它也设置为静态文件目录，而这个目录的访问路径为“/NewFiles”。

例如，文件夹“NewWwwroot”中有一个 test.jpg，那么我们可以通过/NewFiles/test.jpg 的地址来访问它。

15.4 静态文件的授权管理

在默认情况下，静态文件不需要授权，可以公开访问。因为即使采用了授权，UseAuthorization 一般也是写在 UseStaticFiles 后面的，仍然无法生效。

那么如果我们想对其进行授权管理，想到可以改写 StaticFileMiddleware 这个中间件，在其中添加一些自定义的判断条件，但这貌似不够友好；而且这里只能做一些大类的判断，如请求的 IP 地址是否在允许范围内；不支持根据登录用户的权限来判断（如用户只能看到自己上传的图片），因为权限的判断写在这个中间件之后。

这时可以通过 Filter 的方式来处理。在应用目录中新建一个"images"文件夹，但不要把它设置为静态文件目录，这样"images"文件夹在默认情况下是不允许访问的。

通过 Controller 返回文件的方式来处理请求，代码如下：

```
[Route("api/[controller]")]
[AuthorizeFilter]
public class FileController : Controller
{
    [HttpGet("{name}")]
    public FileResult Get(string name)
    {
        var file = Path.Combine(Directory.GetCurrentDirectory(), "images", name);

        return PhysicalFile(file, "application/octet-stream");
    }
}
```

在 AuthorizeFilter 中进行相关判断，代码如下：

```
public class AuthorizeFilter: ActionFilterAttribute
{
    public override void OnActionExecuting(ActionExecutingContext context)
    {
        base.OnActionExecuting(context);

        if (context.RouteData.Values["controller"].ToString().ToLower().Equals("file"))
        {
            bool isAllow = false;//在此进行一系列访问权限验证，如果失败，则返回一个默认图片，
例如 logo 或不允许访问的提示图片

            if (!isAllow)
            {
                var file = Path.Combine(Directory.GetCurrentDirectory(), "images",
"default.png");

                context.Result = new PhysicalFileResult(file,
"application/octet-stream");

            }
        }
    }
}
```

第 16 章　路由

本章对比我们熟悉的 ASP.NET 的 Framework 版本路由，首先讲解路由的配置方法，然后通过一幅图来了解路由的运行机制，最后总结二者的异同点。

16.1 概述

对于 ASP.NET Core 的路由，我个人的理解是：一系列访问 Controller.Action 的规则的集合。它给人的第一印象是如下一条默认设置：

```
app.UseEndpoints(endpoints =>
{
    endpoints.MapControllerRoute(
        name: "default",
        pattern: "{controller=Home}/{action=Index}/{id?}");
});
```

无论是在 ASP.NET Core 还是其 Framework 版本中，都制定了一条访问 Controller.Action 的常用规则，利用从请求的 URL 解析出对应的 Controller.Action，并将请求交给该 Action 处理。

举例来说，将 Controller 比作市、Action 比作区，那么这个模板就可以理解为“去{市=北京市}/{区=朝阳区}/{身份证号}”，这里的“北京市”和“朝阳区”是默认值。例如，说“去天津市和平区，身份证号 *XXX*”，将这句话和模板匹配，分析出目的地是“天津市和平区”，身份证号是“*XXX*”。如果只说了“去”，没有说去哪里，那么会采用默认的地址“北京市朝阳区”。

这条只是默认的规则设置，到达一个地方可以有多种方式，可以乘坐飞机、火车等不同的交通工具，也可以走不同的路线、携带不同的东西，例如身份证、学生证、食品等。所以也可以制定这样的模板：乘飞机去{市=北京市}/{区=朝阳区}/{身份证号}/{行李箱}。

在实际项目中，可以根据项目需求定义多种不同的模板。例如乘飞机，可以添加一条这样的路由：

```
app.UseEndpoints(endpoints =>
{
```

```
    endpoints.MapControllerRoute(
        name: " fly ",
        pattern: " fly/{controller=Home}/{action=Index}/{id}/{bagcode}");
});
```

16.2 传统路由配置

配置路由主要有两种方式，其中一种是在 Startup 中设置，例如默认配置如下：

```
app.UseEndpoints(endpoints =>
{
    endpoints.MapControllerRoute(
        name: "default",
        pattern: "{controller=Home}/{action=Index}/{id?}");
});
```

可以在这个方法中传入一个或多个 endpoints.MapControllerRoute (…)等配置项，我们称之为“传统路由”，它的参数和用法如下。

- **name**：设定了路由的名称。
- **pattern**：路由的模板，其中用“{}”设定了参数的名称，controller 和 action 为保留字，会被自动替换为 controller 和 action 的名称。同时它们被设置了默认值 Home 和 Index，也就是当未提供这两个参数时，会采用这样的默认值。例如，访问 http://localhost:62544 和 http://localhost:62544/home/index 的效果是一样的，就像默认去“北京市朝阳区”。“{id?}”代表这里会接收一个参数，并将其赋值给 id，“？”表示这个参数是可选的。也就是无论访问/home/index 或/home/index/2，都同样会访问到 HomeController 的 Index 这个 Action。如果这个 Action 有参数“id”，第二种访问方式会将参数“id”赋值为 2。
- **defaults**：默认值还可以通过 defaults 参数来设置。例如上面的例子可以改为如下形式：

```
endpoints.MapControllerRoute(name: "default", pattern: "{controller}/{action}/{id?}",
defaults: new { controller = "Home", action = "Index" });
```

看起来比{controller=Home}方式麻烦一些，但如果 template 中没有{controller}参数，defaults 参数的作用就体现出来了。例如如下路由：

```
endpoints.MapControllerRoute(name: "borrow", pattern: "borrow/{**bookid}", defaults:new
{ Controller = "Book",Action="GetById"});
```

此路由设置主要用于一些特定的操作，比如要借书，“borrow/书的 id”要比“book/getById/书的 id”简单且直接。由于设置了默认的 Controller 和 Action，所以通过此路由的 URL 会始终被指向 Book.GetById。这个模板中用到的“**”是一个“贪婪”的匹配符，会将“borrow/”后面的所有字符赋值给变量“bookid”。

16.3 属性路由设置

除了在 Startup 中设置，还可以直接在 Controller 和 Action 中设置，代码如下：

```
public class BookController : Controller
{
    [Route("B/R")]
    public IActionResult Read()
    {
        return new JsonResult($"Read");
    }

    public IActionResult Write()
    {
        return new JsonResult($"Write");
    }

    public IActionResult Drop()
    {
        return new JsonResult($"Drop");
    }

    public IActionResult getById()
    {
        return new JsonResult($"Get");
    }
}
```

对 Action 添加了“[Route("B/R")]”标记，被称为“属性路由”，这样设置后就可以通过 http://localhost:62544/b/r 访问它。路由标记中的“B/R”对应传统路由的 pattern，所以如果 Action 存在参数 bookid，也可以在此添加参数，例如[Route("B/R/{bookid}")]。属性路由同样可以像传统路由一样设置路由的名称，例如[Route("R",Name ="ReadBook")]。现在尝试访问 http://localhost:62544/book/read，发现请求结果是 404 页面。尝试访问没有添加属性路由标记的 http://localhost:62544/book/write，发现是可以正常访问的，这是因为没有对“Write”这个 Action 设置属性路由，它还可以采用传统路由的方式访问。

这里涉及以下 4 个规则。

规则一：“B”和“R”是随意设置的，只是为了方便才采用了首字母，也可以设置成其他的名字。

规则二：当 Action 采用了属性路由后，对应 Action 不再采用传统路由中设置的规则，而没有设置的 Action 仍可采用传统路由中设置的规则。

现在想为“Write”这个 Action 设置属性路由，同时 BookController 下还有多个 Action 需要设置，为了方便，可以将路由的一部分放在 Controller 上，代码如下：

```
[Route("B")]
public class BookController : Controller
{
    [Route("R")]
    public IActionResult Read()
    {
        return new JsonResult("Read");
    }

    [Route("W")]
    public IActionResult Write()
    {
        return new JsonResult("Write");
    }

    public IActionResult Drop(string bookid)
    {
        return new JsonResult($"Drop:{bookid}");
    }

    public IActionResult GetById(string bookid)
    {
         return new JsonResult($"Get:{bookid}");
    }
}
```

这样就可以通过 http://localhost:62544/b/w 访问“Write”这个 Action 了。由于 Controller 和 Action 都可以设置属性路由，因此会涉及多个路由设置之间相互作用的问题。这里涉及另外几条规则。

规则三：如果 Controller 和 Action 都设置了属性路由，系统会自动在 Controller 和 Action 的属性路由值之间添加一个“/”。

规则四：如果 Controller 设置了属性路由，而 Action 没有设置，则该 Action 如同被设置了一个空的属性路由“[Route("")]”。例如本例 BookController 中，没有设置属性路由的名为“Drop”的 Action 只能通过 http://localhost:62544/b 的方式访问。注意，如果有多个这样的 Action，例如本例中还有一个未设置的 Action“GetById”，请求时就会发生“matched multiple endpoints”的错误，因为系统发现有多个 Action 符合要求，不知道最终应将请求交给哪个处理，要尽量避免出现这样的情况。

如果 Controller 没有设置属性路由，而 Action 设置了，见规则二。

属性路由累加：可以累加多条属性路由，无论是 Controller 还是 Action，和上面的设置一样。

```
[Route("MyBook")]
[Route("B")]
public class BookController : Controller
{
    [Route("Re")]
    [Route("R")]
    public IActionResult Read()
    {
        return new JsonResult("Read");
    }
}
```

这样就可以通过多种路由方式访问它，而且 Controller 和 Action 的属性路由不是一一对应的关系，例如可以访问“MyBook/R”，也可以访问“B/R”。Controllcr 上设置的属性路由数量和 Action 上设置的数量可以不一致。

使用 Http[Verb]属性：属性路由还可以使用 Http[Verb]属性，即常用的[HttpGet]、[HttpPost]等，这在 WebAPI 中经常用到。例如，下面的 ValuesController 就是系统默认新增的：

```
[Route("api/[controller]")]
public class ValuesController : Controller
{
    // GET: api/<controller>
    [HttpGet]
    public IEnumerable<string> Get()
    {
        return new string[] { "value1", "value2" };
    }

    // GET api/<controller>/5
    [HttpGet("{id}")]
    public string Get(int id)
    {
        return "value";
    }

    // POST api/<controller>
    [HttpPost]
    public void Post([FromBody]string value)
    {
    }
}
```

这里不但用到了 Http[Verb] 属性，还可以在其中添加参数，例如代码中的“[HttpGet("{id}")]”，这个参数类似传统路由的 template。而 ValuesController 被添加了一个属性路由“[Route("api/[controller]")]”，这里不仅可以采用“B”“R”等固定的自定义名称，也可以添加“controller”“action”等可变参数。所以，对于不同的 WebAPI 的 Controller，它们的属性路由是通用的“[Route("api/[controller]")]”（注意是“[controller]”，而不是“{controller}”），而不需要逐一自定义。那么是否可以通过继承的方式在父类中设置路由，而不用逐一设置呢？答案是可以的。

属性路由的继承：如新建一个 BaseController，并为其设置好属性路由。代码如下：

```
[Route("api/[controller]")]
public class BaseController : Controller
{
}
```

修改上文的 ValuesController，去掉它的属性路由设置，继承 BaseController，代码如下：

```
public class ValuesController : BaseController
{
    //各种 Action
}
```

访问 ValuesController 测试，结果显示和之前一样，访问正常。

16.4 路由的匹配顺序

添加了属性路由的 Action 不再会被传统路由匹配，由此可知，属性路由的优先级大于传统路由。

那么多条传统路由之间的顺序如何呢？下面再添加一条路由，例如：

```
app.UseEndpoints(endpoints => {
    endpoints.MapControllerRoute(name: "default",
            pattern: "{controller}/{action}/{id?}",
            defaults: new { controller = "Home", action = "Index" });
    endpoints.MapControllerRoute(name: "ByName",
            pattern: "{controller=Home}/{action=Index}/{name?}");
});
```

可以看出这两条路由相似，当访问 http://localhost:62544/test/index/1 地址时，都是匹配的，因为“1”既可以赋值给 id，也可以赋值给 name。通过添加一个 TestController 来验证：

```
public class TestController : Controller
{
    public JsonResult Index(string name,string id)
    {
```

```
            return new JsonResult(name + "|" + id);
        }
    }
```

返回的 JSON 文件为“|1”，说明参数 id 被赋值了，采用 name 为 default 的路由。将两条路由互换位置，将新添加的 name 为 ByName 的路由放在上面，再次访问，返回的结果为“1|”。可见，参数 name 被赋值了。这说明路由的匹配依然是按照注册的先后顺序（从代码上看是“从上到下”）进行的，当两个或多个路由设置都能匹配的情况下，先注册的会被采用。

属性路由的顺序见如下代码，为了方便对比，将 BookController 进行修改：

```
public class BookController : Controller
{
    [Route("book/buy/{aa}")]
    [Route("book/buy/{bb}")]
    public IActionResult Buy(string aa,string bb)
    {
        return new JsonResult($"Buy:{aa}|{bb}");
    }
}
```

对 Action 的 Drop 设置了两个几乎一模一样的属性路由，这时会提示错误：

```
AmbiguousMatchException: The request matched multiple endpoints. Matches:
```

说明有多个终结点匹配成功，所以报错。和一个路由匹配多个 Action 出错类似，这次反过来了，多个属性路由匹配了同一个 Action。传统路由可以通过注册的先后顺序来确定类似路由的执行顺序，属性路由不可以吗？从这个例子来看，通过属性路由的“上下顺序”进行区分是不行的。但属性路由有一个 Order 参数，可以设置属性路由的执行顺序，例如：

```
    [Route("book/buy/{aa}",Order = -2)]
    [Route("book/buy/{bb}",Order = -3)]
    public IActionResult Buy(string aa,string bb)
    {
        return new JsonResult($"Buy:{aa}|{bb}");
    }
```

这样设置了 Order 之后就不会报“多对一”的错误了，而且可以通过 Order 的大小来指定执行顺序，不受“上下顺序”的影响。Order 值小的优先级高，例如此例中会匹配参数为“bb”的路由。

对于“一对多”的情况，代码如下：

```
    [Route("book/{**aa}")]
    [Route("book/buy/{bb}")]
    public IActionResult Buy(string aa,string bb)
    {
        return new JsonResult($"Buy:{aa}|{bb}");
    }
```

在这样的路由设置情况下，访问 http://localhost:62544/book/buy/1，理论上这两种路由都是可以匹配的，也没有设置 Order，但这里却不会报错，而且返回值是"Buy:|1"，说明匹配了第二个路由，因为第二个路由更"精确"一些。下面添加 Order 参数：

```
[Route("book/{*aa}",Order = -2)]
[Route("book/buy/{bb}",Order = -1)]
public IActionResult Buy(string aa,string bb)
{
    return new JsonResult($"Buy:{aa}|{bb}");
}
```

返回值依然是"Buy:|1"，依然是第二个路由被采用了，说明路由模板的"精确度"也是影响路由优先级的重要指标。

注意： 通过上面的例子，我们虽然明白了路由的匹配顺序（或者说优先级），但在实际项目中应尽量避免出现这样的路由设置，因为容易混淆，也不易理清关系，在使用和维护上非常容易出错。

16.5 路由的约束

上文讲解路由的顺序时有这样一个例子：

```
app.UseEndpoints(endpoints =>
{
    endpoints.MapControllerRoute(name: "default",
                pattern: "{controller}/{action}/{id?}",
                defaults: new { controller = "Home", action = "Index" });
    endpoints.MapControllerRoute(name: "ByName",
                pattern: "{controller=Home}/{action=Index}/{name?}");
});
```

在此情况下，请求会由先注册的路由匹配。如果原本的想法是让第一个路由来匹配用户的 id，而这个 id 只匹配整数参数，当匹配失败时，由后面的 ByName 路由进行匹配。即当请求为 test/index/1 时，由第一个路由匹配，参数 1 赋值给 id；当请求为 test/index/zhangsan 时，由第二个路由匹配，参数赋值给 name。对于这样的需求，需要对参数 id 进行约束，使其只能匹配整数。下面讲解几种常见的路由约束使用方式。

16.5.1　Constraints 参数方式

对于路由约束，传统路由有一个名为 constraints 的参数，使用 constraints 将参数 id 约束为整数的代码如下：

```
endpoints.MapControllerRoute(name: "default", pattern:
"{controller}/{action}/{id?}",defaults:new { Controller = "Home", Action = "Index" },
constraints:new {id = new IntRouteConstraint() });
```

其中，“IntRouteConstraint”是系统内置的约束，要求 id 的值为 int 型或空。若不符合这样的要求，则对应请求不会被此路由匹配。类似的还有 BoolRouteConstraint、DateTimeRoute Constraint、DoubleRouteConstraint 等，根据名字可以看出作用，就不一一赘述了；复杂一点的有 MaxRouteConstraint(long max) RangeRouteConstraint(long min, long max)等。

可以对多个参数进行约束，多个约束之间也可以嵌套，例如：

```
endpoints.MapControllerRoute(name: "defaultName", pattern:
"{controller}/{action}/{id}/{name}", defaults: new { controller = "Home", action =
"Index" },constraints:new{ id = new IntRouteConstraint(),name = new
CompositeRouteConstraint(new IRouteConstraint[] { new AlphaRouteConstraint(),new
MaxLengthRouteConstraint(10)})});
```

16.5.2 行内简写方式

除了使用 Constraints 的方式，还有一种更简洁的方式：

```
endpoints.MapControllerRoute(name: "default", pattern:
"{controller}/{action}/{id:int?}",defaults:new { Controller = "Home", Action = "Index" });
```

“{id:int?}”使这里只可以接受空或整型的 id 值。例如，访问 http://localhost:62544/test/index/s，则会将 s 赋值给参数 name，对应 MaxRouteConstraint(long max)，这里可以采用行内简写的方式，写成“{id:max(20)}”。

对于属性路由，行内简写约束方式一样适用，例如：

```
 [Route("W/{bookid:maxlength(10)}")]
public IActionResult Write(string bookid)
{
    return new JsonResult($"Write:{bookid}");
}
```

16.5.3 使用正则表达式

对于复杂的约束，可以利用正则表达式实现，例如下面 3 种方式。

方式一：

```
endpoints.MapControllerRoute(name: "default", pattern:
"{controller}/{action}/{id:regex(^([1-9]\\d{{3}})$)}", defaults: new { controller =
"Home", action = "Index" });
```

方式二：

```
[Route("W/{bookid:regex(^([[1-9]]\\d{{3}})$)}")]
public IActionResult Write(string bookid)
{
    return new JsonResult($"Write:{bookid}");
}
```

方式三：

```
endpoints.MapControllerRoute(name: "default", pattern:
"{controller}/{action}/{id}",defaults:new { controller = "Home", action = "Index" },
constraints:new {id = new RegexRouteConstraint("^([1-9]\\d{3})$") });
```

3 种方式的不同点主要是特殊字符转义部分：首先，“\”是非常常见的转义字符，3 种方式均需要被转义为“\\”；其次，对于前两种方式，正则表达式写在了 template 参数中，而这里设置路由的参数需要用“{}”括起来，例如“{id}”，所以这里的“{}”需要被转义为“{{}}”；最后，在属性路由中，Controller 和 Action 等需要表示为“[controller]”和“[action]”，所以这里的“[]”也需要被转义为“[[]]”。注意这些规则不能错误使用，否则该转义的没转义，会报错；不该转义的转义了，会导致该路由匹配失败。

16.5.4 自定义约束

假如有这样一个需求，需要验证某 string 型参数是否满足这样的规则：前两个字符为前缀 XY，总长度为 Z，最终结果如“BJ001”。由于 XY、Z 的值均来自系统设置，如来自 appsettings.json 或数据库。这时采用上文介绍的几种约束方式就无法实现了。参考上文用到的 IntRouteConstraint AlphaRouteConstraint 等内置约束，可以自定义一个类似的约束来实现。

它们都实现了 IRouteConstraint 接口，那么也可以新建一个约束，同样实现这个接口，并将其命名为“ItemCodeRouteConstraint”：

```
    public class ItemCodeRouteConstraint : IRouteConstraint
    {
        public bool Match(HttpContext httpContext, IRouter route, string routeKey,
RouteValueDictionary values, RouteDirection routeDirection)
        {
            //省略各种验证
            //object value;
            if (values.TryGetValue(routeKey,out object value))
            {
                string itemCodePrefix = "BJ"; // 此处只是例子
                int itemCodeLength = 5;          // 实际应从 Configuration 等处获取

                string itemCode = value.ToString();

                return itemCode.StartsWith(itemCodePrefix) && itemCode.Length ==
itemCodeLength;
            }

            return false;
        }
    }
```

再写一个用于测试的 Action：

```
public JsonResult GetItem(string itemCode)
{
    return new JsonResult(itemCode);
}
```

将 ItemCodeRouteConstraint 在 Startup 的 ConfigureServices 中注册使用：

```
services.Configure<RouteOptions>(options=> {
    options.ConstraintMap.Add("ItemCode",typeof( ItemCodeRouteConstraint));
});
```

在传统路由中使用：

```
endpoints.MapControllerRoute(name: "default", pattern:
"{controller}/{action}/{itemcode}",defaults: new { controller = "Home", action = "Index" },
constraints: new { itemcode = new ItemCodeRouteConstraint() });
```

也可以在传统路由和属性路由中的 template 参数中使用，省略 RouteConstraint，直接写 ItemCode，例如：

```
[Route("T/{code:ItemCode}")]
public JsonResult GetItem2(string code)
{
    return new JsonResult(itemCode);
}
```

16.6 路由的 dataTokens

在传统路由中，有一个 dataTokens 参数，它的可用性不高，下面通过它的作用和使用方法来说明。修改默认模板，添加 dataTokens：

```
endpoints.MapControllerRoute(name: "default", pattern:
"{controller=Home}/{action=Index}/{id?}",defaults: null, constraints: null,dataTokens:
new { testTokenName = "testTokenValue" });
```

前面没有做任何改变，从 defaults 参数开始到最后添加的内容。查看 MapRoute 提供的重载，使用 dataTokens 必须提供 defaults 和 constraints，即使赋值为 null 也必须写上。使用如下代码在 Action 中获取它：

```
public JsonResult Index()
{
        string token = RouteData.DataTokens["testTokenName"].ToString();
        return new JsonResult(token);
}
```

需要通过 RouteData 获取，虽然不是很麻烦，但在 Action 中这样获取值给人的感觉“不够友好”。dataTokens 的作用是给所有通过它的请求附加一个 Key-Value 的键值对，所有通过它被访问的 Action 都可以获取到这个值。

根据这个值能知道请求是通过哪个路由访问的。在 MSDN 中有这样一个例子：

```
endpoints.MapControllerRoute(
    name: "us_english_products",
    pattern: "en-US/Products/{id}",
    defaults: new { controller = "Products", action = "Details" },
    constraints: new { id = new IntRouteConstraint() },
    dataTokens: new { locale = "en-US" });
```

大概的意思是标记请求的语言类型为“en-US”，这样系统可以根据这个值返回对应语言的版本。但这样的需求完全可以通过其他方式实现：

```
endpoints.MapControllerRoute(
    name: "products",
    pattern: "{locale}/Products/{id}",
    defaults: new { controller = "Products", action = "Details" },
    constraints: new { id = new IntRouteConstraint() });
```

这样可以在 Action 中通过设置参数 locale 来获取想要的语言类型。对比两种方式，无论是路由的设置，还是值的获取方面，像这样通过路由参数的方式更加方便。注意，通过 dataTokens 将值固定写在传统路由中，不够灵活，路由条数较多时容易隐藏错误。

16.7 路由的初始化源码解析

至此，路由的配置与使用已经讲解得差不多了，下面从代码角度看看路由的内部处理机制，以及路由在应用启动过程中都做了哪些工作。依然是从 Startup 的 Configure 方法开始，路由相关的两个中间件注册是 UseRouting 和 UseEndpoints(……)两个方法。

16.7.1 UseRouting 方法

UseRouting()写在 EndpointRoutingApplicationBuilderExtensions 中，代码如下：

```
public static IApplicationBuilder UseRouting(this IApplicationBuilder builder)
{
    VerifyRoutingServicesAreRegistered(builder);

    var endpointRouteBuilder = new DefaultEndpointRouteBuilder(builder);

    builder.Properties[EndpointRouteBuilder] = endpointRouteBuilder;

    return builder.UseMiddleware<EndpointRoutingMiddleware>(endpointRouteBuilder);
}
```

VerifyRoutingServicesAreRegistered 方法用于验证是否依赖注入容器中已经添加了路由相关的服务。接下来创建一个 DefaultEndpointRouteBuilder，存入 builder.Properties，最后添加中间件 EndpointRoutingMiddleware，新建的 DefaultEndpointRouteBuilder 作为这个中间件的参数。DefaultEndpointRouteBuilder 代码如下：

```
internal class DefaultEndpointRouteBuilder : IEndpointRouteBuilder
{
    public DefaultEndpointRouteBuilder(IApplicationBuilder applicationBuilder)
    {
        ApplicationBuilder = applicationBuilder ?? throw new
ArgumentNullException(nameof(applicationBuilder));
        DataSources = new List<EndpointDataSource>();
    }

    public IApplicationBuilder ApplicationBuilder { get; }

    public IApplicationBuilder CreateApplicationBuilder() => ApplicationBuilder.New();

    public ICollection<EndpointDataSource> DataSources { get; }

    public IServiceProvider ServiceProvider => ApplicationBuilder.ApplicationServices;
}
```

其中有一个重要的成员 ICollection<EndpointDataSource> DataSources，用于存放路由的 EndpointDataSource，后面会用到。DefaultEndpointRouteBuilder 引用了传入的 ApplicationBuilder。

16.7.2 UseEndpoints 方法

UseEndpoints 方法同样写在 EndpointRoutingApplicationBuilderExtensions 中，代码如下：

```
public static IApplicationBuilder UseEndpoints(this IApplicationBuilder builder,
Action<IEndpointRouteBuilder> configure)
{
    VerifyRoutingServicesAreRegistered(builder);

    VerifyEndpointRoutingMiddlewareIsRegistered(builder, out var endpointRouteBuilder);

    configure(endpointRouteBuilder);

    var routeOptions =
builder.ApplicationServices.GetRequiredService<Ioptions<RouteOptions>>();
    foreach (var dataSource in endpointRouteBuilder.DataSources)
    {
        routeOptions.Value.EndpointDataSources.Add(dataSource);
    }
```

```
    return builder.UseMiddleware<EndpointMiddleware>();
}
```

此处多了一条验证，VerifyEndpointRoutingMiddlewareIsRegistered 方法用于验证 UseRouting 方法是否已经被执行，即是否添加了 EndpointRoutingMiddleware 中间件，并且获取在 UseRouting 方法中创建的 DefaultEndpointRouteBuilder，赋值给 endpointRouteBuilder。

configure(endpointRouteBuilder)方法是执行 Startup 中的一个或多个 endpoints.MapControllerRoute()的路由配置方法。此时会创建一个 dataSource 并添加到 DefaultEndpointRouteBuilder 的 DataSources 集合中。而这些 endpoints.MapControllerRoute()路由配置方法会逐条写入 dataSource 的_routes 集合（类型为 List<ConventionalRouteEntry>）。

接下来遍历 endpointRouteBuilder.DataSources，将其中的 dataSource 添加到统一的集合 routeOptions.Value.EndpointDataSources 中，最后添加中间件 EndpointMiddleware。

MapControllerRoute 方法写在 ControllerEndpointRouteBuilderExtensions 中，代码如下：

```
public static ControllerActionEndpointConventionBuilder MapControllerRoute(
    this IEndpointRouteBuilder endpoints,
    string name,
    string pattern,
    object defaults = null,
    object constraints = null,
    object dataTokens = null)
{
    EnsureControllerServices(endpoints);

    var dataSource = GetOrCreateDataSource(endpoints);
    return dataSource.AddRoute(
        name,
        pattern,
        new RouteValueDictionary(defaults),
        new RouteValueDictionary(constraints),
        new RouteValueDictionary(dataTokens));
}
```

首先从依赖注入容器中获取一个 dataSource，并添加到 DefaultEndpointRouteBuilder 的 DataSources 集合中。这个 dataSource 类型为 ControllerActionEndpointDataSource，代码如下（ActionEndpointDataSourceBase 继承自 EndpointDataSource）：

```
internal class ControllerActionEndpointDataSource : ActionEndpointDataSourceBase
{
    private readonly ActionEndpointFactory _endpointFactory;
    private readonly List<ConventionalRouteEntry> _routes;
    public ControllerActionEndpointConventionBuilder AddRoute(
        string routeName,
```

```
        string pattern,
        RouteValueDictionary defaults,
        IDictionary<string, object> constraints,
        RouteValueDictionary dataTokens)
    {
        lock (Lock)
        {
            var conventions = new List<Action<EndpointBuilder>>();
            _routes.Add(new ConventionalRouteEntry(routeName, pattern, defaults,
constraints, dataTokens, _order++, conventions));
            return new ControllerActionEndpointConventionBuilder(Lock, conventions);
        }
}

    //此处省略了 CreateEndpoints 方法，下文用到时会讲

}
```

Startup 中配置的一条或多条 endpoints.MapControllerRoute(……)通过调用 ControllerAction EndpointDataSource 的 AddRoute 方法逐一转换成 ConventionalRouteEntry，并写入 dataSource 的_routes 集合（类型为 List<ConventionalRouteEntry>）。

图 16-1 所示为 UseRouting 和 UseEndpoints 方法的两个过程。

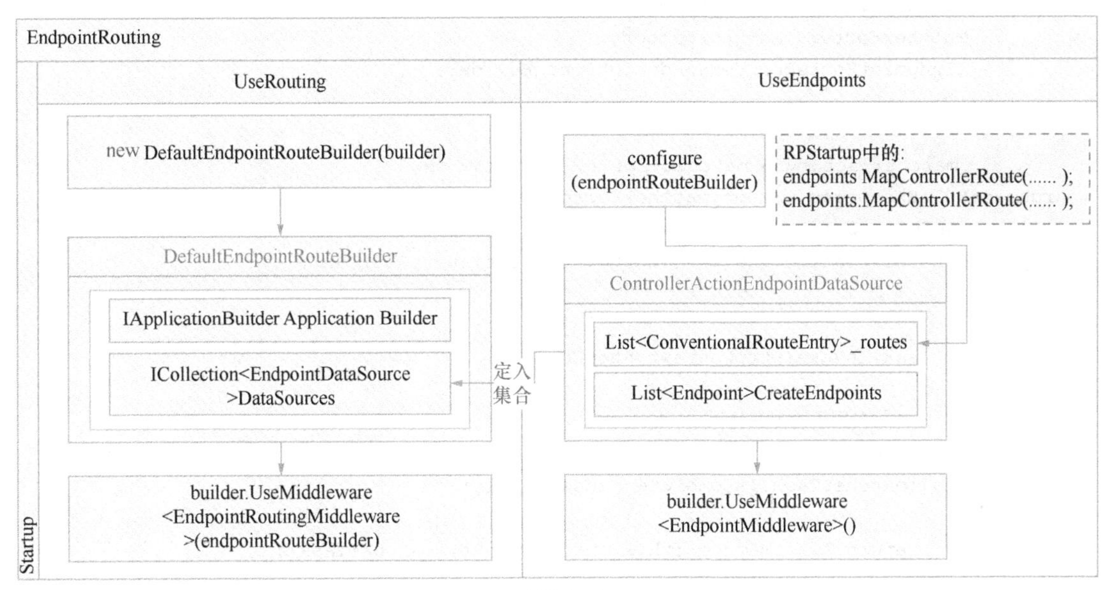

▲图 16-1

16.8 路由的请求处理源码分析

上一节讲了应用启动阶段添加了路由相关的两个中间件 EndpointRoutingMiddleware 和 EndpointMiddleware，本节讲解应用接收到请求时，路由是如何工作的。

请求处理的大部分功能在中间件 EndpointRoutingMiddleware 中实现，属性 _endpointDataSource 保存了初始化阶段生成的 MvcEndpointDataSource。而中间件 EndpointMiddleware 的功能比较简单，主要是在 EndpointRoutingMiddleware 筛选出 endpoint 之后，调用该 endpoint 的 endpoint.RequestDelegate(httpContext)进行请求处理。

16.8.1 EndpointRoutingMiddleware

EndpointRoutingMiddleware 简化后的代码如下：

```
internal sealed class EndpointRoutingMiddleware
{

    private readonly MatcherFactory _matcherFactory;
    private readonly EndpointDataSource _endpointDataSource;
    private Task<Matcher> _initializationTask;

    public EndpointRoutingMiddleware(
        MatcherFactory matcherFactory,
        IEndpointRouteBuilder endpointRouteBuilder,
    )
    {
        _matcherFactory = matcherFactory ?? throw new
ArgumentNullException(nameof(matcherFactory));
        _endpointDataSource = new
CompositeEndpointDataSource(endpointRouteBuilder.DataSources);
    }

    public Task Invoke(HttpContext httpContext)
    {
        // 省略部分用于验证的代码
        var matcherTask = InitializeAsync();
        if (!matcherTask.IsCompletedSuccessfully)
        {
            return AwaitMatcher(this, httpContext, matcherTask);
        }

        var matchTask = matcherTask.Result.MatchAsync(httpContext);
        if (!matchTask.IsCompletedSuccessfully)
        {
```

```
            return AwaitMatch(this, httpContext, matchTask);
        }

        return SetRoutingAndContinue(httpContext);

        static async Task AwaitMatcher(EndpointRoutingMiddleware middleware,
HttpContext httpContext, Task<Matcher> matcherTask)
        {
            var matcher = await matcherTask;
            await matcher.MatchAsync(httpContext);
            await middleware.SetRoutingAndContinue(httpContext);
        }

        static async Task AwaitMatch(EndpointRoutingMiddleware middleware,
HttpContext httpContext, Task matchTask)
        {
            await matchTask;
            await middleware.SetRoutingAndContinue(httpContext);
        }

    }
    private Task<Matcher> InitializeAsync()
    {
        var initializationTask = _initializationTask;
        if (initializationTask != null)
        {
            return initializationTask;
        }

        return InitializeCoreAsync();
    }

    private Task<Matcher> InitializeCoreAsync()
    {
        var initialization = new
TaskCompletionSource<Matcher>(TaskCreationOptions.RunContinuationsAsynchronously);
        var initializationTask = Interlocked.CompareExchange(ref _initializationTask,
 initialization.Task, null);
        if (initializationTask != null)
        {
            return initializationTask;
        }

        try
        {
            var matcher = _matcherFactory.CreateMatcher(_endpointDataSource);
```

```
            using (ExecutionContext.SuppressFlow())
            {
                _initializationTask = Task.FromResult(matcher);
            }

            initialization.SetResult(matcher);
            return initialization.Task;
        }
        catch (Exception ex)
        {
        }
    }
}
```

由图 16-1 可知，在 UseRouting 方法的最后添加 EndpointRoutingMiddleware 时，将 DefaultEndpointRouteBuilder 作为参数。所以在 EndpointRoutingMiddleware 的构造方法中获取了传入的 DefaultEndpointRouteBuilder，此时的 DefaultEndpointRouteBuilder 的 DataSources 集合中已经携带 UseEndpoints 方法中获取的路由配置，也就是 ControllerActionEndpointDataSource，这是因为 UseEndpoints 方法的执行在 EndpointRoutingMiddleware 创建之前。EndpointRoutingMiddleware 在它的构造方法中将 DefaultEndpointRouteBuilder 的 DataSources 集合进一步封装成了_endpointDataSource，它属于 CompositeEndpointDataSource 类型，所以此时 EndpointRoutingMiddleware 的_endpointDataSource 集合包含了 ControllerActionEndpointDataSource，即路由配置信息。

请求的处理发生在 EndpointRoutingMiddleware 的 Invoke 方法中，调用 InitializeAsync()创建一个 Matcher，类型为 DfaMatcher。从 Matcher 的名字可知，它的作用是实现路由的匹配。InitializeAsync 方法主要是通过调用 InitializeCoreAsync()创建 Matcher 的，通过这个方法的代码可以看出，它只在第一次请求时执行一次，而生产 Matcher 的工作交给了 InitializeCoreAsync。

```
private Task<Matcher> InitializeAsync()
{
    var initializationTask = _initializationTask;
    if (initializationTask != null)
    {
    return initializationTask;
    }

    return InitializeCoreAsync();
}
```

InitializeCoreAsync 的主要工作是调用_matcherFactory，即调用 DfaMatcherFactory 的 Matcher CreateMatcher(EndpointDataSource dataSource)方法创建 Matcher。

```
var matcher = _matcherFactory.CreateMatcher(_endpointDataSource);
```

既然 Matcher 用于路由的匹配，那么必然会用到路由配置，这里的参数用到了 EndpointRoutingMiddleware 的_endpointDataSource 集合，核心内容实际上是图 16-1 中的 ControllerActionEndpointDataSource。

ControllerActionEndpointDataSource 的一个重要方法是 CreateEndpoints，创建 Matcher 的 CreateMatcher(_endpointDataSource)方法时，会逐步调用这个方法。它的作用是读取所有 Action，将所有 Action 逐一与 ControllerActionEndpointDataSource 的_routes 集合（传统路由）以及 Action 自身的属性路由进行匹配，最终生成一个 Endpoints 的列表。具体代码如下：

```
internal class ControllerActionEndpointDataSource : ActionEndpointDataSourceBase
{
    private readonly ActionEndpointFactory _endpointFactory;
    private readonly List<ConventionalRouteEntry> _routes;

    protected override List<Endpoint> CreateEndpoints(IReadOnlyList<ActionDescriptor>
 actions, IReadOnlyList<Action<EndpointBuilder>> conventions)
    {
        var endpoints = new List<Endpoint>();
        var keys = new HashSet<string>(StringComparer.OrdinalIgnoreCase);

        var routeNames = new HashSet<string>(StringComparer.OrdinalIgnoreCase);
        for (var i = 0; i < actions.Count; i++)
        {
            if (actions[i] is ControllerActionDescriptor action)
            {
                _endpointFactory.AddEndpoints(endpoints, routeNames, action, _routes,
 conventions, CreateInertEndpoints);

                if (_routes.Count > 0)
                {
                    foreach (var kvp in action.RouteValues)
                    {
                        keys.Add(kvp.Key);
                    }
                }
            }
        }

        for (var i = 0; i < _routes.Count; i++)
        {
            var route = _routes[i];
            _endpointFactory.AddConventionalLinkGenerationRoute(endpoints, routeNames,
keys, route, conventions);
        }
```

```
        return endpoints;
    }
}
```

核心操作是_endpointFactory.AddEndpoints(endpoints, routeNames, action, _routes, conventions, CreateInertEndpoints)方法，写在 ActionEndpointFactory 中，精简后的代码如下：

```
internal class ActionEndpointFactory
{
    public void AddEndpoints(
        List<Endpoint> endpoints,
        HashSet<string> routeNames,
        ActionDescriptor action,
        IReadOnlyList<ConventionalRouteEntry> routes,
        IReadOnlyList<Action<EndpointBuilder>> conventions,
        bool createInertEndpoints)
    {
        if (action.AttributeRouteInfo == null)
        {
            foreach (var route in routes)
            {
                var updatedRoutePattern =
_routePatternTransformer.SubstituteRequiredValues(route.Pattern, action.RouteValues);
                if (updatedRoutePattern == null)
                {
                    continue;
                }
                var builder = new RouteEndpointBuilder(_requestDelegate, updatedRoute
Pattern, route.Order)
                {
                    DisplayName = action.DisplayName,
                };
                AddActionDataToBuilder(
                    builder,
                    routeNames,
                    action,
                    route.RouteName,
                    route.DataTokens,
                    suppressLinkGeneration: true,
                    suppressPathMatching: false,
                    conventions,
                    route.Conventions);
                endpoints.Add(builder.Build());
            }
        }
```

```
        else
        {
            var attributeRoutePattern =
RoutePatternFactory.Parse(action.AttributeRouteInfo.Template);

            var (resolvedRoutePattern, resolvedRouteValues) =
ResolveDefaultsAndRequiredValues(action, attributeRoutePattern);

            var updatedRoutePattern =
_routePatternTransformer.SubstituteRequiredValues(resolvedRoutePattern, resolvedRouteValues);
            if (updatedRoutePattern == null)
            {
                throw new InvalidOperationException("Failed to update route pattern
with required values.");
            }

            var builder = new RouteEndpointBuilder( requestDelegate, updatedRoute
Pattern, action.AttributeRouteInfo.Order)
            {
                DisplayName = action.DisplayName,
            };
            AddActionDataToBuilder(
                builder,
                routeNames,
                action,
                action.AttributeRouteInfo.Name,
                dataTokens: null,
                action.AttributeRouteInfo.SuppressLinkGeneration,
                action.AttributeRouteInfo.SuppressPathMatching,
                conventions,
                perRouteConventions: Array.Empty<Action<EndpointBuilder>>());
            endpoints.Add(builder.Build());
        }
    }
}
```

判断当前 Action 是否被设置了属性路由，如果是，则按照属性路由创建 Endpoint；否则与 Startup 中设置的传统路由做匹配，生成相应的 Endpoint。所以 CreateEndpoints 方法的本质是计算出一个个可能被请求的请求终结点，也就是 Endpoint。生成的 Endpoint 集合会被进一步处理，进行切片、排序（根据传统路由的先后顺序以及属性路由的 Order 参数），最终生成一个路由匹配规则的集合 DfaState[]。

Matcher 被创建后，调用 Matcher 的 Task MatchAsync(HttpContext httpContext)方法，将请求 URL 与处理好的路由规则做匹配。第一次匹配会进行一次初步的筛选，生成一个候选集合 Candidates，代码如下：

```
internal (Candidate[] candidates, IEndpointSelectorPolicy[] policies) FindCandidateSet(
    HttpContext httpContext,
    string path,
    ReadOnlySpan<PathSegment> segments)
{
    var states = _states;
    var destination = 0;
    for (var i = 0; i < segments.Length; i++)
    {
        destination = states[destination].PathTransitions.GetDestination(path,
segments[i]);
    }
    var policyTransitions = states[destination].PolicyTransitions;
    while (policyTransitions != null)
    {
        destination = policyTransitions.GetDestination(httpContext);
        policyTransitions = states[destination].PolicyTransitions;
    }

    return (states[destination].Candidates, states[destination].Policies);
}
```

接下来遍历 Candidates 集合，找到最佳的 Endpoint，这里会判断备选项的一些配置，例如默认值、约束等，最终确定最佳的 Endpoint，并赋值给 feature.Endpoint。约束的判断方法如下：

```
private bool ProcessConstraints(Endpoint endpoint,
    KeyValuePair<string, IRouteConstraint>[] constraints,
    HttpContext httpContext, RouteValueDictionary values)
{
    for (var i = 0; i < constraints.Length; i++)
    {
        var constraint = constraints[i];
        if (!constraint.Value.Match(httpContext, NullRouter.Instance, constraint.Key,
 values, RouteDirection.IncomingRequest))
        {
            Logger.CandidateRejectedByConstraint(_logger, httpContext.Request.Path,
endpoint, constraint.Key, constraint.Value, values[constraint.Key]);
            return false;
        }
    }

    return true;
}
```

至此，我们找到了最佳的 Endpoint，EndpointRoutingMiddleware 中间件的工作完成。

16.8.2 Endpoint 的生成与匹配示例

为了更好地理解，下面以路由配置为例来演示这个过程。在 Startup 中设置传统路由：

```
endpoints.MapControllerRoute(name: "default",
                    pattern: "{controller=Home}/{action=Index}/{id?}");
endpoints.MapControllerRoute(name: "name",
                    pattern: "{controller}/{action}/{name?}");
```

HomeController 中保留两个默认的 Action：

```
public class HomeController : Controller
{
    public IActionResult Index()
    {
        return View();
    }

    public IActionResult Privacy()
    {
        return View();
    }
}
```

加上前面的例子 BookController：

```
public class BookController : Controller
{
    [Route("B/R")]
    public IActionResult Read()
    {
        return new JsonResult("Read");
    }

    public IActionResult Write()
    {
        return new JsonResult("Write");
    }



    [Route("book/buy/{aa}", Order = -1)]
    [Route("book/buy/{bb}", Order = -2)]
    public IActionResult Buy(string aa, string bb)
    {
        return new JsonResult($"Drop:{aa}|{bb}");
    }
}
```

以这样的配置来看路由的处理和匹配的过程，如图 16-2 所示。这里只用默认程序举一个简单的例子，实际项目中可能会有更多的路由模板注册，也会有更多的 Controller 和 Action 以及属性路由等。

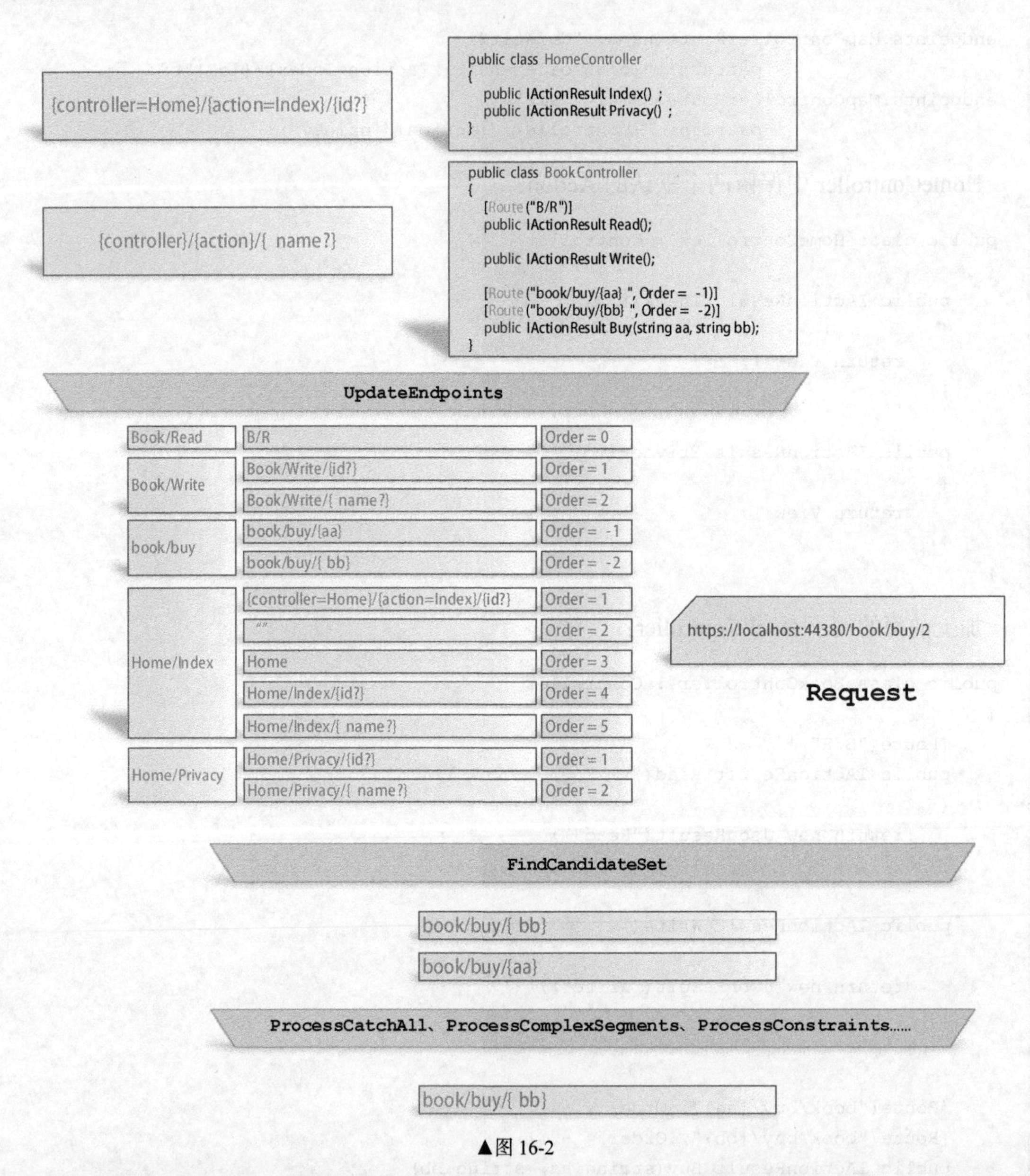

▲图 16-2

在这样的配置情况下，5 个 Action 与两个传统路由以及几个属性路由通过 UpdateEndpoints 方法做匹配，有点像“乘积”，生成了 12 个 Endpoint。当收到 https://localhost:44380/book/buy/2 这样的请求后，二者通过 FindCandidateSet 进行初步筛选，选择 book/buy/{bb}和 book/buy/{aa}

两个备选项，注意这时顺序已经生效了。进一步对这两个备选项进行验证，例如路由约束在此处是否生效，最终选出 book/buy/{bb}为最佳 Endpoint。

16.8.3 EndpointMiddleware

EndpointMiddleware 的 Invoke 方法简化后的代码如下：

```
internal sealed class EndpointMiddleware
{
    public Task Invoke(HttpContext httpContext)
    {
        var endpoint = httpContext.GetEndpoint();
        if (endpoint?.RequestDelegate != null)
        {
            try
            {
                var requestTask = endpoint.RequestDelegate(httpContext);
                if (!requestTask.IsCompletedSuccessfully)
                {
                    return AwaitRequestTask(endpoint, requestTask, _logger);
                }
            }
            return Task.CompletedTask;
        }

        return _next(httpContext);
    }
}
```

核心操作是获取 EndpointRoutingMiddleware 中间件匹配的 Endpoint，并调用其 RequestDelegate 方法来处理请求上下文的 httpContext。

16.9 Endpoint 模式的路由方案的优点

Endpoint 模式的路由方案是从 ASP.NET Core 2.2 开始使用的。下面从应用系统启动和请求处理两个阶段对比 ASP.NET Core 2.2 前后两个路由方案版本的区别（2.1 版本及之前版本与 Framework 版本原理上类似）。

1. 启动阶段

启动阶段大部分是在 Startup 中配置一个路由表，生成一个 Routes 集合，然后将其简单转换。

2.1 版本及之前的版本（包括 2.1 版本）：将 Routes 转换为 RouteCollection。

2.2 版本以后的版本：将 Routes 转换为 List<ConventionalRouteEntry>。

二者虽然名字不同，但本质相似，都可以理解为 Routes 集合的封装。

2. 请求处理阶段

2.1 版本及之前的版本：将请求的 URL 与 RouteCollection 中记录的路由模板进行匹配；找到匹配的 Route 后，再根据这个请求的 URL 判断是否存在对应的 Controller 和 Action；若以上均通过，则调用 Route 的 Handler 对 httpContext 进行处理。

2.2 版本以后的版本：第一次处理请求时，根据启动阶段所配置的路由集合 List<ConventionalRouteEntry>和所有的 Action 做匹配，生成一个列表。这个列表存储了所有可能被匹配请求的 URL 模板，实际上是列出了一个个可以被访问的详细地址，算是最终地址，即终结点；请求的 URL 和这个生成的表做匹配，找到对应的 Endpoint；调用被匹配的 Endpoint 的 RequestDelegate 方法对请求进行处理。

二者的区别在于对 URL 的匹配上。2.1 版本及以前的版本是先根据路由模板匹配后，再根据 ActionDescriptors 判断是否存在对应的 Controller 和 Action，而 2.2 版本以后的版本是先利用 Action 信息与路由模板匹配，然后用请求的 URL 进行匹配。由于这样的工作只在第一次请求时执行，所以虽然没有测试执行效率，但应该是比之前快的。

第 17 章　Action 的执行

上一章介绍了 ASP.NET Core 的路由，一个请求经过路由处理后，找到了具体处理这个请求的 Endpoint，并最终执行它的 RequestDelegate 方法来处理 httpContext。本章继续利用这个处理讲程，来说明 RequestDelegate 方法。

17.1 概述

在 EndpointMiddleware 中间件中，一个请求经过路由的处理找到了对应的 Endpoint，并最终将 httpContext 交给 endpoint 的 RequestDelegate 方法来处理。

请求最终要访问具体的 Controller 和 Action，所以接下来的工作是根据找到的 Endpoint 创建对应的 Controller，并执行对应的 Action。

假如我们现在访问了 Home/Index 这个默认的 Action：

```
public class HomeController : Controller
{
    public IActionResult Index()
    {
        return View();
}
}
```

以这个请求为例，下面通过一幅图来看这个过程，如图 17-1 所示。

17.2 invoker 的生成

在图 17-1 中，每个泳道相当于上一个泳道中的□□□图标的细化说明，例如第二条泳道是图标①标识的方块的细化说明，也就是说，从上到下的各个泳道是逐级递进的关系。

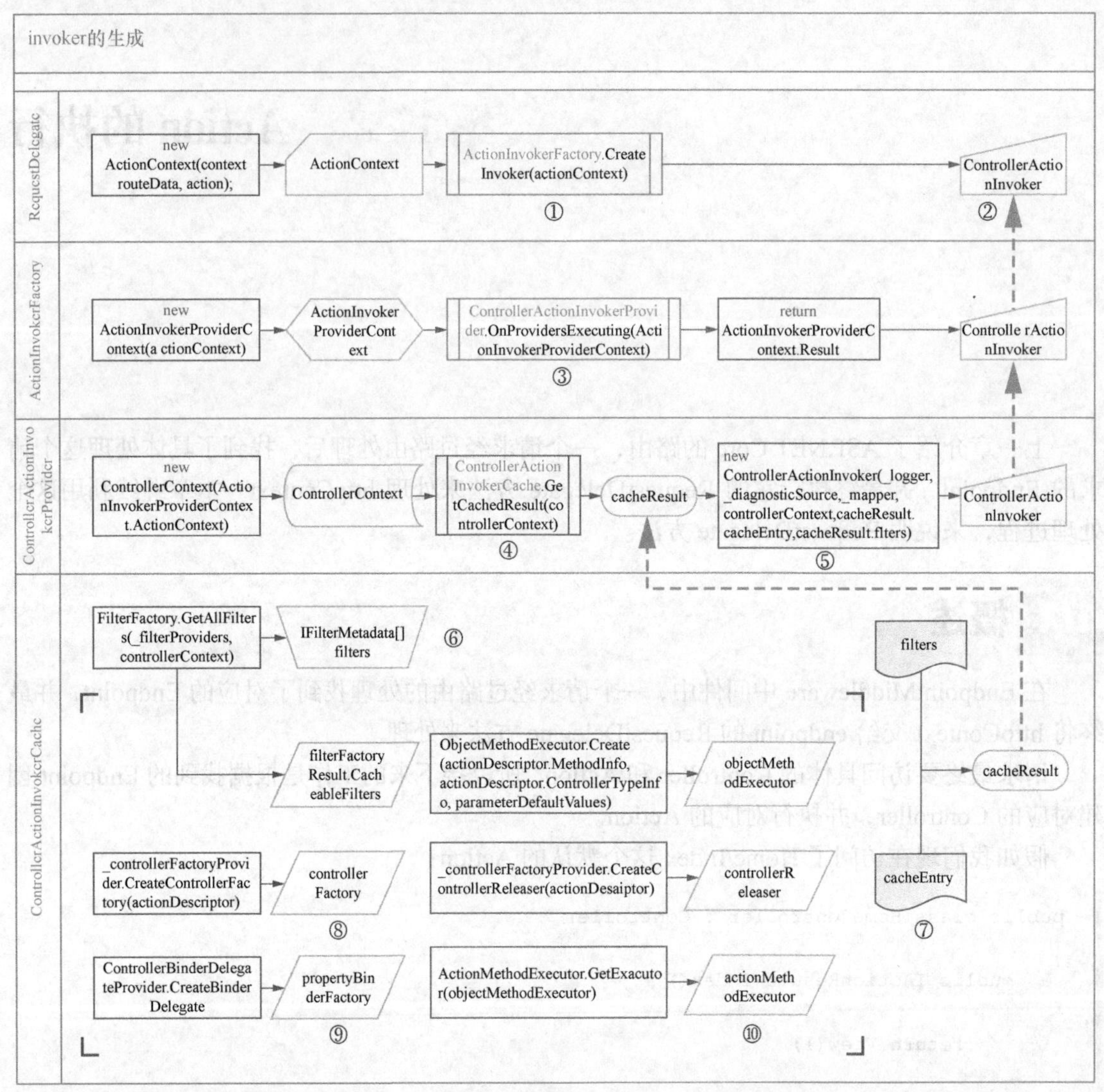

▲图 17-1

泳道一：即 Endpoint 的 RequestDelegate 方法，被写在 ActionEndpointFactory 类中。

```
private static RequestDelegate CreateRequestDelegate()
{
    IActionInvokerFactory invokerFactory = null;
    return (context) =>
    {
        var endpoint = context.GetEndpoint();
        var dataTokens = endpoint.Metadata.GetMetadata<IDataTokensMetadata>();
```

```
        var routeData = new RouteData();
        routeData.PushState(router: null, context.Request.RouteValues, new RouteValue
Dictionary(dataTokens?.DataTokens));

        var action = endpoint.Metadata.GetMetadata<ActionDescriptor>();
        var actionContext = new ActionContext(context, routeData, action);

        if (invokerFactory == null)
        {
            invokerFactory = context.RequestServices.GetRequiredService<IActionInvoke
rFactory>();
        }

        var invoker = invokerFactory.CreateInvoker(actionContext);
        return invoker.InvokeAsync();
    };
}
```

将这个方法的内容分为两部分：invoker 的生成和 invoker 的执行。invoker 是本节的核心，其本质上是一个 ControllerActionInvoker，即图 17-1 中的 ActionInvokerProviderContext.Result。

- invoker 的生成：利用路由的 Action 信息、routeData 和 dataTokens 等，通过 CreateInvoker 方法生成一个 invoker。它是一个比较复杂的综合体，包含 Controller 的创建工厂、参数的绑定方法以及本 Action 相关的各种 Filter 的集合等。也就是说，它属于前期准备工作阶段。
- invoker 的执行：此时涉及各种方法的执行，各种 Filter 也在此时被执行。

泳道二：即对①ActionInvokerFactory.CreateInvoker(actionContext)的详细描述。相关代码如下：

```
internal class ActionInvokerFactory : IActionInvokerFactory
{
    private readonly IActionInvokerProvider[] _actionInvokerProviders;

    public ActionInvokerFactory(IEnumerable<IActionInvokerProvider>
actionInvokerProviders)
    {
        _actionInvokerProviders = actionInvokerProviders.OrderBy(item => item.Order).
ToArray();
    }

    public IActionInvoker CreateInvoker(ActionContext actionContext)
    {
        var context = new ActionInvokerProviderContext(actionContext);

        foreach (var provider in _actionInvokerProviders)
```

```
            {
                provider.OnProvidersExecuting(context);
            }

            for (var i = _actionInvokerProviders.Length - 1; i >= 0; i--)
            {
                _actionInvokerProviders[i].OnProvidersExecuted(context);
            }

            return context.Result;
        }
    }
```

这部分内容中比较常见的一个操作是对 httpContext 的封装，这从第一个泳道的第一个操作就开始了，它将 httpContext、routeData、ActionDescriptor 封装到一起，成为一个 actionContext。在第二个泳道，又将 actionContext 封装成 ActionInvokerProviderContext。接下来遍历_actionInvokerProviders，调用它们的 OnProvidersExecuting 和 OnProvidersExecuted 方法来设置 ActionInvokerProviderContext.Result，也就是最终的②ControllerActionInvoker。

接下来会用到一个数组_actionInvokerProviders，它的类型是 IActionInvokerProvider[]。接口 IActionInvokerProvider 的代码如下：

```
public interface IActionInvokerProvider
{
    int Order { get; }
    void OnProvidersExecuting(ActionInvokerProviderContext context);
    void OnProvidersExecuted(ActionInvokerProviderContext context);
}
```

_actionInvokerProviders 默认包含一个子项 ControllerActionInvokerProvider，用于 MVC 的 Action 的处理（ASP.NET Core 2.0 还包含了一个 PageActionInvokerProvider，用于 Razor Pages Web 的处理，然后动态判断是否执行）。由 IActionInvokerProvider 可知，ControllerActionInvoker Provider 存在一个 OnProvidersExecuting 方法和一个 OnProvidersExecuted 方法。ControllerActionInvoker Provider 的 OnProvidersExecuted 方法目前均为空，所以重点看 ControllerActionInvokerProvider 的 OnProvidersExecuting 方法，即③，详细描述见泳道三。

泳道三：ControllerActionInvokerProvider.OnProvidersExecuting(ActionInvokerProviderContext)，即对泳道二中③的详细描述。相关代码如下：

```
public void OnProvidersExecuting(ActionInvokerProviderContext context)
{
    if (context == null)
    {
        throw new ArgumentNullException(nameof(context));
    }
```

```
    if (context.ActionContext.ActionDescriptor is ControllerActionDescriptor)
    {
        var controllerContext = new ControllerContext(context.ActionContext)
        {
            ValueProviderFactories = new CopyOnWriteList<IValueProviderFactory>
(_valueProviderFactories)
        };
        controllerContext.ModelState.MaxAllowedErrors = _maxModelValidationErrors;

        var (cacheEntry, filters) = _controllerActionInvokerCache.GetCachedResult
(controllerContext);

        var invoker = new ControllerActionInvoker(
            _logger,
            _diagnosticListener,
            _actionContextAccessor,
            _mapper,
            controllerContext,
            cacheEntry,
            filters);

        context.Result = invoker;
    }
}
```

在处理之前，判断当前Action是否是对应处理的类型，是ControllerAction类型还是Razor Pages类型。然后将actionContext封装成ControllerContext，进而调用GetCachedResult方法读取两个关键内容cacheResult.cacheEntry和cacheResult.filters后，将其封装成⑤ControllerActionInvoker。

泳道四：对应泳道三中的④ControllerActionInvokerCache.GetCachedResult(controller Context)，invoker的许多重要组件都在这里组装。泳道四的树状结构本质上就是invoker的组成结构，相关代码如下：

```
 1 public (ControllerActionInvokerCacheEntry cacheEntry, IFilterMetadata[] filters)
GetCachedResult(ControllerContext controllerContext)
 2 {
 3     var cache = CurrentCache;
 4     var actionDescriptor = controllerContext.ActionDescriptor;
 5
 6     IFilterMetadata[] filters;
 7     if (!cache.Entries.TryGetValue(actionDescriptor, out var cacheEntry))
 8     {
 9         var filterFactoryResult = FilterFactory.GetAllFilters(_filterProviders,
controllerContext);
```

```
10          filters = filterFactoryResult.Filters;
11
12          var parameterDefaultValues = ParameterDefaultValues
13              .GetParameterDefaultValues(actionDescriptor.MethodInfo);
14
15          var objectMethodExecutor = ObjectMethodExecutor.Create(
16              actionDescriptor.MethodInfo,
17              actionDescriptor.ControllerTypeInfo,
18              parameterDefaultValues);
19
20          var controllerFactory = _controllerFactoryProvider.CreateControllerFactory
(actionDescriptor);
21          var controllerReleaser = _controllerFactoryProvider.CreateControllerReleaser
(actionDescriptor);
22          var propertyBinderFactory = ControllerBinderDelegateProvider.CreateBinder
Delegate(
23              _parameterBinder,
24              _modelBinderFactory,
25              _modelMetadataProvider,
26              actionDescriptor,
27              _mvcOptions);
28
29          var actionMethodExecutor = ActionMethodExecutor.GetExecutor(objectMethod
Executor);
30
31          cacheEntry = new ControllerActionInvokerCacheEntry(
32              filterFactoryResult.CacheableFilters,
33              controllerFactory,
34              controllerReleaser,
35              propertyBinderFactory,
36              objectMethodExecutor,
37              actionMethodExecutor);
38          cacheEntry = cache.Entries.GetOrAdd(actionDescriptor, cacheEntry);
39      }
40      else
41      {
42          // Filter instances from statically defined filter descriptors + from
filter providers
43          filters = FilterFactory.CreateUncachedFilters(_filterProviders,
controllerContext, cacheEntry.CachedFilters);
44      }
45
46      return (cacheEntry, filters);
```

总体来看，本段内容主要是为了组装 cacheEntry 和 filters 两个内容。从第 7 行的 if 可以看出，这里加入了缓存机制，使系统不必每次都去拼凑这些内容，从而提高执行效率。

⑥IFilterMetadata[] filters 是一个 Filter 的集合，首先调用 FilterFactory 的 GetAllFilters (_filterProviders，controllerContext)方法获取当前 Action 对应的所有 Filter，并对这些 Filter 进行排序。

接下来是组装⑦cacheEntry，其中比较重要的内容如下。

⑧ controllerFactory 和 controllerReleaser 的本质都是 Func<ControllerContext，object>，也就是 Controller 的 Create 和 Release 方法，例如，CreateControllerFactory 会返回一个 CreateController 方法用于创建 Controller。相关代码如下：

```
public Func<ControllerContext, object> CreateControllerFactory(ControllerActionDescriptor
descriptor)
{
    if (descriptor == null)
    {
        throw new ArgumentNullException(nameof(descriptor));
    }

    var controllerType = descriptor.ControllerTypeInfo?.AsType();
    if (controllerType == null)
    {
        throw new ArgumentException(Resources.FormatPropertyOfTypeCannotBeNull(
            nameof(descriptor.ControllerTypeInfo),
            nameof(descriptor)),
            nameof(descriptor));
    }

    if (_factoryCreateController != null)
    {
        return _factoryCreateController;
    }

    var controllerActivator = _activatorProvider.CreateActivator(descriptor);
    var propertyActivators = GetPropertiesToActivate(descriptor);
    object CreateController(ControllerContext controllerContext)
    {
        var controller = controllerActivator(controllerContext);
        for (var i = 0; i < propertyActivators.Length; i++)
        {
            var propertyActivator = propertyActivators[i];
            propertyActivator(controllerContext, controller);
        }
        return controller;
}

    return CreateController;
}
```

⑨propertyBinderFactory 是一个用于参数绑定的 Task，也是一个组装好的准备被执行的方法，作用是 Action 的参数绑定。

最后，⑩actionMethodExecutor 是执行者，对于不同返回值类型的 Action，需要对应的 Executor 来处理。系统默认定义了 8 个这样的 Executor，如图 17-2 所示，这里通过 ActionMethodExecutor.GetExecutor(objectMethodExecutor)方法从众多的 Executor 中找出一个当前 Action 对应的执行者。

```
[0] {Microsoft.AspNetCore.Mvc.Internal.ActionMethodExecutor.VoidResultExecutor}
[1] {Microsoft.AspNetCore.Mvc.Internal.ActionMethodExecutor.SyncActionResultExecutor}
[2] {Microsoft.AspNetCore.Mvc.Internal.ActionMethodExecutor.SyncObjectResultExecutor}
[3] {Microsoft.AspNetCore.Mvc.Internal.ActionMethodExecutor.AwaitableResultExecutor}
[4] {Microsoft.AspNetCore.Mvc.Internal.ActionMethodExecutor.TaskResultExecutor}
[5] {Microsoft.AspNetCore.Mvc.Internal.ActionMethodExecutor.TaskOfIActionResultExecutor}
[6] {Microsoft.AspNetCore.Mvc.Internal.ActionMethodExecutor.TaskOfActionResultExecutor}
[7] {Microsoft.AspNetCore.Mvc.Internal.ActionMethodExecutor.AwaitableObjectResultExecutor}
```

▲图 17-2

通过它们的名字可以大概看出其作用，对不同类型的 Action 的执行者 Executor，均写在名为 ActionMethodExecutor 的类中。它们都继承自 ActionMethodExecutor，有一个 CanExecute 方法，以 VoidResultExecutor 为例，其代码如下：

```
private class VoidResultExecutor : ActionMethodExecutor
{
    public override ValueTask<IActionResult> Execute(
        IActionResultTypeMapper mapper,
        ObjectMethodExecutor executor,
        object controller,
        object[] arguments)
    {
        executor.Execute(controller, arguments);
        return new ValueTask<IActionResult>(new EmptyResult());
    }

    protected override bool CanExecute(ObjectMethodExecutor executor)
        => !executor.IsMethodAsync && executor.MethodReturnType == typeof(void);
}
```

CanExecute 方法是判断依据，遍历这些 executor，通过其 CanExecute 方法进行判断，最终选择出一个适合的 executor，代码如下：

```
public static ActionMethodExecutor GetExecutor(ObjectMethodExecutor executor)
{
    for (var i = 0; i < Executors.Length; i++)
    {
        if (Executors[i].CanExecute(executor))
```

```
            {
                return Executors[i];
            }
        }

        Debug.Fail("Should not get here");
        throw new Exception();
    }
```

至此，invoker 的创建工作基本完成了。

总结： invoker 的生成可以说是一个执行前“万事俱备”的过程。invoker 是一个组装起来的集合，包含一个人（执行者 ActionMethodExecutor）、N 把枪（组装好用于“被执行”的方法，例如 controllerFactory、controllerReleaser 和 propertyBinderFactory，还有一个 filters 集合）。由此可以进一步想到，接下来的过程是按照一定的顺序逐步执行这些准备好的内容。

17.3 invoker 的执行

创建好 invoker 之后，下一步是 invoker 的执行，也就是 invoker.InvokeAsync 方法。虽然 invoker 本质上是 ControllerActionInvoker，但这个方法写在 ResourceInvoker 中，它们的继承关系是“ControllerActionInvoker : ResourceInvoker, IActionInvoker”。

```
public virtual async Task InvokeAsync()
{
    try
    {
        await InvokeFilterPipelineAsync();
    }
    //省略部分代码
}
```

这是一个比较简单的方法，会调用另一个 Task InvokeFilterPipelineAsync 方法。借用 MSDN 中的图，如图 17-3 所示。

图 17-3 描述了请求经过其他中间件处理后，进入路由处理，最终找到对应的 Action，最终进入 Filter 管道进行处理。

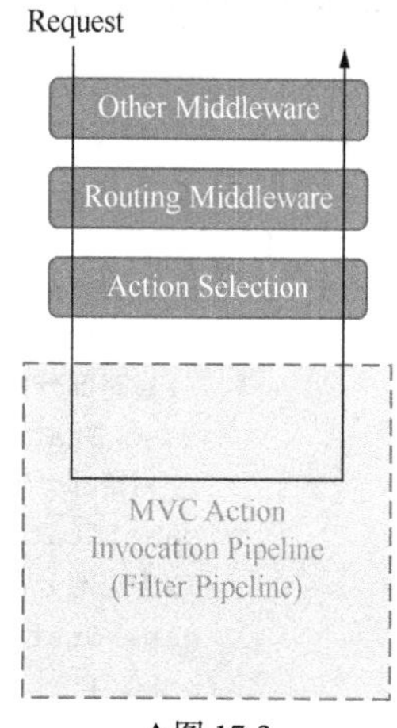

▲图 17-3

InvokeFilterPipelineAsync 方法的代码如下：

```
private Task InvokeFilterPipelineAsync()
{
    var next = State.InvokeBegin;
    var scope = Scope.Invoker;

    var state = (object?)null;

    var isCompleted = false;
```

```
    try
    {
        while (!isCompleted)
        {
            var lastTask = Next(ref next, ref scope, ref state, ref isCompleted);
            if (!lastTask.IsCompletedSuccessfully)
            {
                return Awaited(this, lastTask, next, scope, state, isCompleted);
            }
        }

        return Task.CompletedTask;
    }
    catch (Exception ex)
    {
        return Task.FromException(ex);
    }

    static async Task Awaited(ResourceInvoker invoker, Task lastTask, State next,
Scope scope, object? state, bool isCompleted)
    {
        await lastTask;

        while (!isCompleted)
        {
            await invoker.Next(ref next, ref scope, ref state, ref isCompleted);
        }
    }
}
```

这个方法的核心部分是 while (!isCompleted) 循环，它会根据 State 的状态循环调用 Next 方法。对应的 Next 方法比较长，为了方便描述，将其缩减，只保留了部分 case 的内容，如下（需要完整版的读者请自行查看 ResourceInvoker 的方法）：

```
private Task Next(ref State next, ref Scope scope, ref object state, ref bool isCompleted)
{
    switch (next)
    {
        case State.InvokeBegin:
        case State.AuthorizationBegin:
        //省略部分代码
        case State.ResourceBegin:
        //省略部分代码
        case State.ExceptionBegin:
        //省略部分代码
        case State.ActionBegin:
            {
```

```
                var task = InvokeInnerFilterAsync();
                if (task.Status != TaskStatus.RanToCompletion)
                {
                    next = State.ActionEnd;
                    return task;
                }

                goto case State.ActionEnd;
            }
        case State.ActionEnd:
        case State.ResourceInsideEnd:
        case State.ResourceEnd:
        case State.InvokeEnd:

        default:
            throw new InvalidOperationException();
    }
}
```

此方法中各个 case 的先后顺序大体遵循实际的运行顺序。从代码中可以看出，根据状态 State 进行轮转，大概顺序是 Authorization→Resource→Exception……这里是 FilterPipeline，其作用是按照一定的顺序来执行目标 Action 对应的 Filter，而执行顺序也是 AuthorizationFilter→ResourceFilter→ExceptionFilter……

ActionFilter 的执行比较特殊，它将 Action 的执行放在了中间，也就是在 Action 的执行前后都有 ActionFilter 的逻辑，这和中间件相似。ActionFilter 和 Action 的执行逻辑写在了 ControllerActionInvoker 中，同样是一个 Task Next 方法被 while 循环调用，精简后的代码如下：

```
private Task Next(ref State next, ref Scope scope, ref object state, ref bool isCompleted)
{
    switch (next)
    {
        case State.ActionBegin:
            {
                //省略部分代码
                _arguments = new Dictionary<string, object>(StringComparer.
OrdinalIgnoreCase);

                var task = BindArgumentsAsync();
                if (task.Status != TaskStatus.RanToCompletion)
                {
                    next = State.ActionNext;
                    return task;
                }
```

```
                goto case State.ActionNext;
            }

        case State.ActionNext:
        case State.ActionAsyncBegin:
        case State.ActionAsyncEnd:
        case State.ActionSyncBegin:
        case State.ActionSyncEnd:
        case State.ActionInside:
            {
                var task = InvokeActionMethodAsync();
                if (task.Status != TaskStatus.RanToCompletion)
                {
                    next = State.ActionEnd;
                    return task;
                }

                goto case State.ActionEnd;
            }

        case State.ActionEnd:

        default:
            throw new InvalidOperationException();
    }
}
```

在执行 ActionBegin 的时候，通过 ControllerFactory 创建了 Controller，然后调用 cacheEntry.ControllerBinderDelegate(_controllerContext, _instance, _arguments) 进行参数绑定。

在 State.ActionInside 时，调用 Action 的执行方法，会调用 InvokeActionMethodAsync 方法来执行 Action 内的代码。其他几个 case 的内容是执行 Action 的各个 ActionFilter。它们的顺序是 ActionFilter 的 OnActionExecuting 方法、Action 的执行（即 InvokeActionMethodAsync）、ActionFilter 的 OnActionExecuted 方法。Filter 的相关内容将在第 19 章中做详细描述。

InvokeActionMethodAsync 精简后的代码如下：

```
private Task InvokeActionMethodAsync()
{
    var objectMethodExecutor = _cacheEntry.ObjectMethodExecutor;
    var actionMethodExecutor = _cacheEntry.ActionMethodExecutor;
    var orderedArguments = PrepareArguments(_arguments, objectMethodExecutor);

    var actionResultValueTask = actionMethodExecutor.Execute(_mapper,
objectMethodExecutor,_instance, orderedArguments);
```

```
    if (actionResultValueTask.IsCompletedSuccessfully)
    {
        _result = actionResultValueTask.Result;
    }
    else
    {
        return Awaited(this, actionResultValueTask);
    }

    return Task.CompletedTask;

    static async Task Awaited(ControllerActionInvoker invoker, ValueTask<IActionResult>
actionResultValueTask)
    {
        invoker._result = await actionResultValueTask;
    }
}
```

InvokeActionMethodAsync 方法的作用是利用 actionMethodExecutor 来执行 Action，即 actionMethodExecutor.Execute(_mapper, objectMethodExecutor, _instance, orderedArguments)。actionMethodExecutor 是从几种系统内置的 ActionMethodExecutor 中根据 Action 的类型筛选出来的，此部分内容见图 17-2 及其前后部分。

例如访问默认的 Home/Index 时，对应的是 SyncActionResultExecutor，代码如下：

```
    private class SyncActionResultExecutor : ActionMethodExecutor
    {
        public override ValueTask<IActionResult> Execute(
            IActionResultTypeMapper mapper,
            ObjectMethodExecutor executor,
            object controller,
            object[] arguments)
        {
            var actionResult = (IActionResult)executor.Execute(controller, arguments);
            EnsureActionResultNotNull(executor, actionResult);

            return new ValueTask<IActionResult>(actionResult);
        }

        protected override bool CanExecute(ObjectMethodExecutor executor)
            => !executor.IsMethodAsync &&
typeof(IActionResult).IsAssignableFrom(executor.MethodReturnType);
    }
```

最终，SyncActionResultExecutor 完成了对 Action 方法的执行。

总结：本节的内容是将准备阶段组装的多个方法按一定的顺序逐步执行的过程。关键代码是两个 Next 方法，看上去很复杂，读者只需从宏观上知道这两个 Next 方法的作用是执行目标 Action 对应的各个 Filter、执行 Action 方法，以及将以上两条按照一定的顺序调度执行。

从微观角度来看，在执行过程中还涉及许多内容，例如参数绑定、Filter 的执行顺序、Action 执行后的结果返回等，后面会以单独的章节详细讲解这些内容。

第 18 章　Action 参数的模型绑定

在 Action 的激活中，一个关键的操作是 Action 参数的模型绑定，既涉及简单的 string、int 等类型，也包含 JSON 等复杂类型，本章将详细介绍这一过程。

18.1 概述

当客户端发出一个请求时，参数可能存在于 URL 中，也可能在请求的 Body 中。参数类型也大不相同，可能是简单类型的参数，如字符串、整数或浮点数，也可能是复杂类型的参数，如常见的 JSON、XML 等。这些如何与目标 Action 的参数关联在一起并赋值呢？

例如，存在如下 Action：

```
[HttpPost]
public JsonResult Post([FromBody]Book book,string note)
{
    book.Code = note + "|" + Guid.NewGuid().ToString();
    return new JsonResult(book);
}
```

其中有两个不同类型的参数，一个参数 book 为 Book 类型，对应一个实体类；另一个参数是简单的 string 类型的 note。

```
public class Book
{
    public string Code { get; set; }
    public string Name { get; set; }
}
```

当客户端请求这个 Action 时，我们一般用 JSON 的数据格式发送 Book 实体类型。例如，用 Fiddler 做一个 POST 请求，如图 18-1 所示，两个参数分别通过 Body 和 URL 的方式发送。

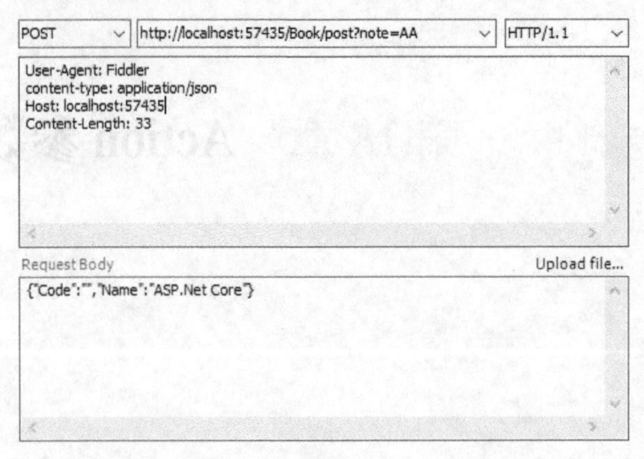

▲图 18-1

得到如图 18-2 所示的请求结果。

▲图 18-2

可见在 Action 中，我们能直接使用两个参数的值，系统已经自动将 Request Body 中的 JSON 转换为 Book 类型，并赋值给了两个对应的参数。这发生在通过路由确定了被请求的 Action 之后，invoker 的创建与执行阶段内容见 18.3 节。在 invoker 的创建与执行阶段，涉及的内容如下。

- invoker 的创建阶段：创建处理方法，并根据目标 Action 的 ActionDescriptor 获取它的所有参数，分析各个参数的类型，并确定对应参数的绑定方法。
- invoker 的执行阶段：调用处理方法，遍历参数逐一进行赋值。

下面依然以这个请求为例，来了解系统对其处理过程。

18.2 准备阶段

在 invoker 的创建阶段会创建一些关键对象和用于执行的方法，本章需要讲解的对象和方法也包含在其中。

18.2.1　创建绑定方法

当收到请求后，由路由系统确定被访问的目标 Action 是我们定义的 Action——Post。这时进入 invoker 的创建阶段，invoker 的关键属性之一 CacheEntry 由多个对象组装而成（发生在 ControllerActionInvokerCache 的 GetCachedResult 方法中）。propertyBinderFactory 是众多方法中的一个，它是一个用于参数绑定的 Task。

propertyBinderFactory 的创建过程如下：

```
var propertyBinderFactory = ControllerBinderDelegateProvider.CreateBinderDelegate
(_parameterBinder,_modelBinderFactory,_modelMetadataProvider,actionDescriptor,_mvcOptions);
```

这个方法被定义在 ControllerBinderDelegateProvider 中，其代码如下：

```
public    static    ControllerBinderDelegate    CreateBinderDelegate(ParameterBinder
parameterBinder,IModelBinderFactory modelBinderFactory,
             IModelMetadataProvider modelMetadataProvider, ControllerActionDescriptor
 actionDescriptor,  MvcOptions mvcOptions)
{
    //省略各种验证

    var parameterBindingInfo = GetParameterBindingInfo(modelBinderFactory,
modelMetadataProvider, actionDescriptor, mvcOptions);
    var propertyBindingInfo = GetPropertyBindingInfo(modelBinderFactory,
modelMetadataProvider, actionDescriptor);
    if (parameterBindingInfo == null && propertyBindingInfo == null)
    {
        return null;
    }
    return Bind;

    async Task Bind(ControllerContext controllerContext, object controller,
Dictionary<string, object> arguments)
    {
        //后文详细描述
    }
 }
```

可见其本质上是一个名为 Bind 的 Task，作为 invoker 的一部分等待被执行。

18.2.2 为每个参数匹配 Binder

CreateBinderDelegate 方法创建了两个对象：parameterBindingInfo 和 propertyBindingInfo。顾名思义，前者用于参数绑定，后者用于属性绑定，二者实现方式类似。这里以参数绑定为例，来看 parameterBindingInfo 的创建：

```
private static BinderItem[] GetParameterBindingInfo(IModelBinderFactory
modelBinderFactory,IModelMetadataProvider, modelMetadataProvider,
ControllerActionDescriptor actionDescriptor, MvcOptions mvcOptions)
        {
            var parameters = actionDescriptor.Parameters;
            if (parameters.Count == 0)
            {
                return null;
```

```
            }
            var parameterBindingInfo = new BinderItem[parameters.Count];
            for (var i = 0; i < parameters.Count; i++)
            {
                var parameter = parameters[i];
                   //省略部分代码
                var binder = modelBinderFactory.CreateBinder(new ModelBinderFactory
Context
                {
                    BindingInfo = parameter.BindingInfo,
                    Metadata = metadata,
                    CacheToken = parameter,
                });

                parameterBindingInfo[i] = new BinderItem(binder, metadata);
            }

            return parameterBindingInfo;
        }
```

可以看到，parameterBindingInfo 本质是一个 BinderItem[]，通过遍历目标 Action 的所有参数 actionDescriptor.Parameters，为每个参数逐一匹配一个对应的处理对象 BinderItem。

本例中会匹配两个 Binder。

- book 参数：{Microsoft.AspNetCore.Mvc.ModelBinding.Binders.BodyModelBinder}。
- note 参数：{Microsoft.AspNetCore.Mvc.ModelBinding.Binders.SimpleTypeModelBinder}。

```
private readonly struct BinderItem
{
    public BinderItem(IModelBinder modelBinder, ModelMetadata modelMetadata)
    {
        ModelBinder = modelBinder;
         ModelMetadata = modelMetadata;
    }

    public IModelBinder ModelBinder { get; }

    public ModelMetadata ModelMetadata { get; }
}
```

两个不同的参数匹配了两个不同的 BinderItem，这是因为系统定义了一系列 Provider，如图 18-3 所示。

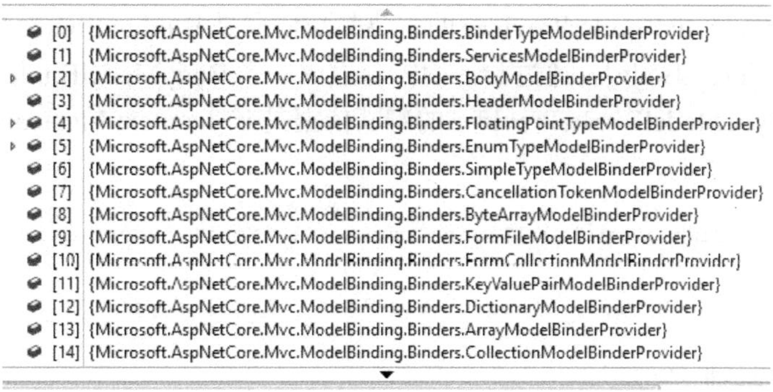

```
[0] {Microsoft.AspNetCore.Mvc.ModelBinding.Binders.BinderTypeModelBinderProvider}
[1] {Microsoft.AspNetCore.Mvc.ModelBinding.Binders.ServicesModelBinderProvider}
[2] {Microsoft.AspNetCore.Mvc.ModelBinding.Binders.BodyModelBinderProvider}
[3] {Microsoft.AspNetCore.Mvc.ModelBinding.Binders.HeaderModelBinderProvider}
[4] {Microsoft.AspNetCore.Mvc.ModelBinding.Binders.FloatingPointTypeModelBinderProvider}
[5] {Microsoft.AspNetCore.Mvc.ModelBinding.Binders.EnumTypeModelBinderProvider}
[6] {Microsoft.AspNetCore.Mvc.ModelBinding.Binders.SimpleTypeModelBinderProvider}
[7] {Microsoft.AspNetCore.Mvc.ModelBinding.Binders.CancellationTokenModelBinderProvider}
[8] {Microsoft.AspNetCore.Mvc.ModelBinding.Binders.ByteArrayModelBinderProvider}
[9] {Microsoft.AspNetCore.Mvc.ModelBinding.Binders.FormFileModelBinderProvider}
[10] {Microsoft.AspNetCore.Mvc.ModelBinding.Binders.FormCollectionModelBinderProvider}
[11] {Microsoft.AspNetCore.Mvc.ModelBinding.Binders.KeyValuePairModelBinderProvider}
[12] {Microsoft.AspNetCore.Mvc.ModelBinding.Binders.DictionaryModelBinderProvider}
[13] {Microsoft.AspNetCore.Mvc.ModelBinding.Binders.ArrayModelBinderProvider}
[14] {Microsoft.AspNetCore.Mvc.ModelBinding.Binders.CollectionModelBinderProvider}
```

▲图 18-3

系统会使它们逐一与当前参数进行匹配，当第一次匹配成功后，则不再与后面的 Provider 进行匹配，代码如下。

```
for (var i = 0; i < _providers.Length; i++)
{
    var provider = _providers[i];
    result = provider.GetBinder(providerContext);
    if (result != null)
    {
        break;
    }
}
```

同样以这两个参数对应的 Binder 为例，BodyModelBinderProvider 的判断代码如下：

```
public IModelBinder GetBinder(ModelBinderProviderContext context)
{
    //省略非空验证
    if (context.BindingInfo.BindingSource != null &&
        context.BindingInfo.BindingSource.CanAcceptDataFrom(BindingSource.Body))
    {
        if (_formatters.Count == 0)
        {
            throw new InvalidOperationException(Resources.FormatInput
FormattersAreRequired(
                typeof(MvcOptions).FullName,
                nameof(MvcOptions.InputFormatters),
                typeof(IInputFormatter).FullName));
        }

        return new BodyModelBinder(_formatters, _readerFactory, _loggerFactory,
_options);
    }

    return null;
}
```

可以看到，BodyModelBinderProvider 的主要判断依据是 BindingSource.Body，也就是因为 book 参数被设置了[FromBody]。通过验证后，会生成一个 BodyModelBinder 并返回。

同理，SimpleTypeModelBinderProvider 的判断依据是 if (!context.Metadata.IsComplexType)，代码如下：

```
public class SimpleTypeModelBinderProvider : IModelBinderProvider
{
    public IModelBinder GetBinder(ModelBinderProviderContext context)
    {
        //省略非空验证
        if (!context.Metadata.IsComplexType)
        {
            var loggerFactory = context.Services.GetRequiredService<ILoggerFactory>();
            return new SimpleTypeModelBinder(context.Metadata.ModelType, loggerFactory);
        }

        return null;
    }
}
```

通过上述的匹配方式为每个参数找到对应的 Provider，则会由该 Provider 生成一个对应的 ModelBinder 并返回，也就有了前面的 BodyModelBinder 和 SimpleTypeModelBinder。

总结：至此前期准备工作已经完成，这里创建了 3 个重要的对象。

- **Task Bind()**：用于绑定的方法，并被封装到 invoker 内的 CacheEntry 中。
- **parameterBindingInfo**：本质是一个 BinderItem[]，其中的 BinderItem 数量与 Action 的参数数量相同。
- **propertyBindingInfo**：类似 parameterBindingInfo，用于属性绑定。

三者的隶属结构如图 18-4 所示。

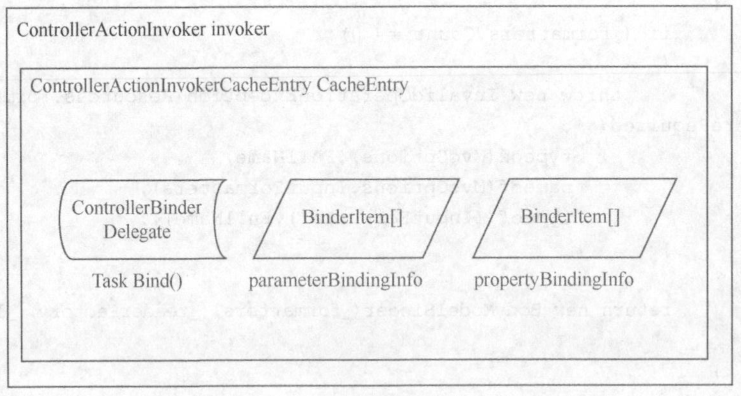

▲图 18-4

18.3 执行阶段

从 18.2 节的总结可以联想到，执行阶段会调用 Bind 方法，利用创建的 parameterBindingInfo 和 propertyBindingInfo 将请求发送的参数处理后赋值给 Action 对应的参数。

执行阶段发生在 invoker（即 ControllerActionInvoker）的 InvokeAsync 方法中，它有一个由 State 控制的 Next 方法。当调用它的 Next 方法时，State 为 ActionBegin 时就会调用 BindArgumentsAsync 方法，如下：

```
private Task Next(ref State next, ref Scope scope, ref object state, ref bool isCompleted)
{
    switch (next)
    {
        case State.ActionBegin:
        {
         //省略部分代码
           _arguments = new Dictionary<string, object>(StringComparer.OrdinalIgnoreCase);

           var task = BindArgumentsAsync();
        }
         //省略部分代码
    }
}
```

而 BindArgumentsAsync 方法会调用 18.2 节中创建的_cacheEntry.ControllerBinderDelegate，也就是 Task Bind 方法。代码如下：

```
private Task BindArgumentsAsync()
{
    //省略部分代码
    return _cacheEntry.ControllerBinderDelegate(_controllerContext, _instance,
_arguments);
}
```

18.2 节省略了 Bind 方法，下面详细来看这个方法：

```
async Task Bind(ControllerContext controllerContext, object controller, Dictionary
<string, object> arguments)
{
    var valueProvider = await CompositeValueProvider.CreateAsync(controllerContext);
    var parameters = actionDescriptor.Parameters;

    for (var i = 0; i < parameters.Count; i++) //遍历参数集合，逐一处理
    {
       var parameter = parameters[i];
```

```
        var bindingInfo = parameterBindingInfo[i];
        var modelMetadata = bindingInfo.ModelMetadata;

        if (!modelMetadata.IsBindingAllowed)
        {
            continue;
        }

        var result = await parameterBinder.BindModelAsync(
                controllerContext,bindingInfo.ModelBinder,
                valueProvider, parameter,  modelMetadata, value: null);

         if (result.IsModelSet)
         {
             arguments[parameter.Name] = result.Model;
         }
    }

    var properties = actionDescriptor.BoundProperties;
    for (var i = 0; i < properties.Count; i++)
        //省略部分代码
    }
}
```

主体是两个 for 循环，分别用于处理参数和属性，以参数处理为例进行说明。

首先获取 Action 所有的参数，然后进入 for 循环进行遍历，通过 parameterBindingInfo[i]获取参数对应的 BinderItem，这些都准备好后，调用 parameterBinder.BindModelAsync 方法进行参数处理和赋值。注意，这里传入了 bindingInfo.ModelBinder，也就是上文获取的 BodyModelBinder 和 SimpleTypeModelBinder，在 BindModelAsync 中会调用传入的 ModelBinder 的 BindModelAsync 方法 modelBinder.BindModelAsync(modelBindingContext)。现在已经将被处理对象交给了 BodyModelBinder、SimpleTypeModelBinder 等具体的 ModelBinder 了。

以 BodyModelBinder 为例：

```
public async Task BindModelAsync(ModelBindingContext bindingContext)
{
    //省略部分代码
    var formatterContext = new InputFormatterContext(httpContext,modelBindingKey,
bindingContext.ModelState, bindingContext.ModelMetadata,  _readerFactory, allowEmpty
InputInModelBinding);
    var formatter = (IInputFormatter)null;
    for (var i = 0; i < _formatters.Count; i++)
  {
        if (_formatters[i].CanRead(formatterContext))
```

```
            {
                formatter = _formatters[i];
                _logger?.InputFormatterSelected(formatter, formatterContext);
                break;
            }
            else
            {
                _logger?.InputFormatterRejected(_formatters[i], formatterContext);
            }
        }

        var result = await formatter.ReadAsync(formatterContext);
        //省略部分代码
    }
```

部分代码已省略，从已有代码中可以看到，这里像上文匹配 Provider 一样，会遍历一个名为_formatters 的集合，通过子项的 CanRead 方法来确定是否可以处理 formatterContext。若可以，则调用该 formatter 的 ReadAsync 方法进行处理。_formatters 集合默认有一个 formatter，是 Microsoft.AspNetCore.Mvc.Formatters.SystemTextJsonInputFormatter。

在 ASP.NET Core 2 中，采用 Newtonsoft.Json 作为默认处理程序。默认有两个 formatter，分别是 JsonInputFormatter 和 JsonPatchInputFormatter。JsonPatchInputFormatter 的判断逻辑如下：

```
if (!typeof(IJsonPatchDocument).GetTypeInfo().IsAssignableFrom(modelTypeInfo) ||
            !modelTypeInfo.IsGenericType)
{
    return false;
}
```

它会判断请求的类型是否为 IJsonPatchDocument（JsonPatch 见 18.4 节）。我们经常使用的是 JsonInputFormatter，此处采用 Newtonsoft.Json 则会匹配 JsonInputFormatter。

本例默认采用 System.Text.Json 处理 JSON 文件，匹配到 SystemTextJsonInputFormatter，它继承自 TextInputFormatter，TextInputFormatter 又继承自 InputFormatter。SystemTextJsonInputFormatter 未重写 CanRead 方法，而是采用 InputFormatter 的 CanRead 方法。

```
public virtual bool CanRead(InputFormatterContext context)
{
    if (SupportedMediaTypes.Count == 0)
    {
        var message = Resources.FormatFormatter_NoMediaTypes(GetType().FullName,
nameof(SupportedMediaTypes));
        throw new InvalidOperationException(message);
    }
```

```
    if (!CanReadType(context.ModelType))
    {
        return false;
    }

    var contentType = context.HttpContext.Request.ContentType;
    if (string.IsNullOrEmpty(contentType))
    {
        return false;
    }
    return IsSubsetOfAnySupportedContentType(contentType);
}
```

要求 Content-Type 不能为空。本例的参数为[FromBody]Book book，并标识了 content-type: application/json。通过 CanRead 方法验证后，则调用 ReadRequestBodyAsync 方法处理 Body 中的 JSON：

```
public sealed override async Task<InputFormatterResult> ReadRequestBodyAsync(
    InputFormatterContext context,
    Encoding encoding)
{

    var httpContext = context.HttpContext;
    var inputStream = GetInputStream(httpContext, encoding);

    object model;
    try
    {
        model = await JsonSerializer.DeserializeAsync(inputStream, context.ModelType,
SerializerOptions);
    }

    //省略部分代码
    return InputFormatterResult.Success(model);}
```

可以看到此处是将收到的请求内容 Deserialize 赋给 model，再返回获取。

总结：本阶段的工作是获取请求参数的值并赋值给 Action 的对应参数。由于参数不同，会分配到一些不同的处理方法中处理。例如本例涉及的 Provider、不同的 ModelBinder（BodyModelBinder 和 SimpleTypeModelBinder）、不同的 formatter 等，实际项目中还会遇到其他的类型，这里不再赘述。

18.4 相关知识

上一节中有两个需要单独说明的参数，在这里简单介绍。

18.4.1 propertyBindingInfo

propertyBindingInfo 主要用于处理 Controller 的属性的赋值，例如：

```
public class BookController : Controller
{
    [ModelBinder]
    public string Key { get; set; }
```

属性 Key 被标记为[ModelBinder]，会在 Action 被请求时被赋值。修改上面的例子，对该 URL 添加一个参数&key=ss。再次请求，会发现这个 Key 像参数一样被赋值为 ss，其处理规则和 Action 的参数处理规则类似，在此不再赘述。

18.4.2 JsonPatch

JsonPatch 可以被理解为操作 JSON 的文档，比如上文的 User 类如下：

```
public class Book
{
    public string Code { get; set; }
    public string Name { get; set; }
    //省略部分代码
}
```

现在如果只修改它的 Name 属性，默认情况下仍然需要提交这样的 JSON：

```
{"Code":"001","Name":"ASP", ... }
```

但这不科学！从节省流量的角度来说，参数也太多了，用 JsonPatch 可以写为：

```
[
  { "op" : "replace", "path" : "/Name", "value" : "ASP" }
]
```

第 19 章　Filter 详解

Filter 在系统中经常被用到，本章详细介绍各种 Filter 定义、执行的内部机制，以及执行顺序。

19.1 概述

在讲解 Action 的执行时涉及了 FilterPipeLine，即筛选器管道。筛选器管道可以说是 ASP.NET Core 的请求处理管道的一部分，因为会按照一定的顺序执行被请求的 Action 对应的各个 Filter（即筛选器，下文统称为 Filter）。Filter 的作用是在 Action 被执行前后进行一些特殊的检查及处理。

如果将请求处理管道理解为一条公路，那么 FilterPipeLine 就是这条公路上的一段特殊路段，在这段路上预留了几个用于设置检查站或服务区的位置。Filter 就是这一特殊路段上的各个检查站或服务区。在默认情况下，有的位置可能是空的，有的位置会存在一个或几个系统默认设置的检查站或服务区。例如 Authorization filters，即授权筛选器，就像公路上用作安检的检查站，作用是判断当前的请求是否已被授权。如果请求未被授权，则终止执行后面的处理请求，所以它一般会放在前面执行。

其他几种筛选器如下。

（1）**Resource filters**：资源筛选器，在授权后运行。例如处理缓存，可以在授权后直接返回缓存结果而不进入后面的筛选器处理。它有如下两个方法。

- **OnResourceExecuting**：一般在授权筛选器执行之后执行。
- **OnResourceExecuted**：一般在筛选器管道的最后执行。

（2）**Action filters**：操作筛选器，在 Action 的前后执行，例如 Action 执行前的参数处理和 Action 执行后对 Action 的返回结果的处理。

（3）**Exception filters**：异常筛选器，用于在向响应正文写入任何内容之前，对未经处理的异常应用进行全局的异常处理策略。

（4）**Result filters**：结果筛选器，在请求结果被执行的前后执行。仅当操作方法成功执行时，它们才会运行。

在实际项目中，根据这些筛选器的作用以及项目需求，可以自定义所需要类型的 Filter，并将其注册到 FilterPipeLine 中。

本章对 Filter 从定义到执行的 4 个阶段进行说明，如图 19-1 所示。

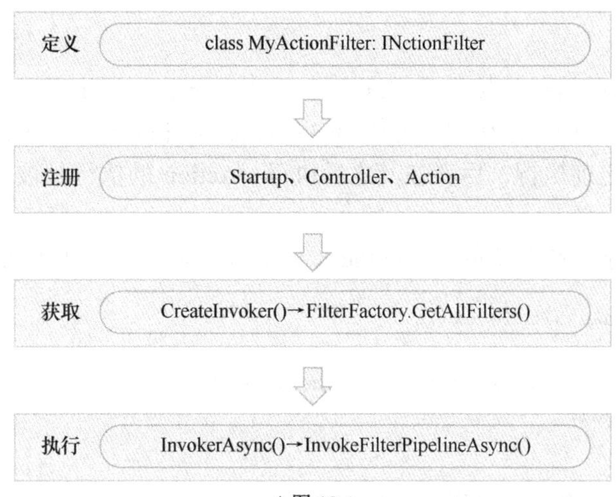

▲图 19-1

（1）定义：这里以 ActionFilter 为例，可以通过继承 ActionFilterAttribute 并覆盖它的 OnActionExecuting 和 OnActionExecuted 方法实现。

（2）注册：主要有 3 种方式，可以在 Startup、Controller 或 Action 中注册。

（3）获取：在确定了处理请求的 Endpoint 后，下一步是创建 invoker，它由 FilterFactory 的 GetAllFilters 方法获取，并有一个关键的属性 filters。

（4）执行：invoker 的执行阶段会调用 InvokeFilterPipelineAsync 方法。在这里，各种 Filter 会按照图 19-1 的方式逐一被执行。

本章首先通过一个例子简要说明如何自定义并使用一个 Filter，然后通过代码详细讲解 Filter 的执行过程、执行顺序等。

19.2 Filter 的简单例子

在 19.1 节中简单介绍了多种 Filter，但定义与使用方法相似。由于本文主要介绍 Filter 的使用与运行机制，所以此处只以 ActionFilter 为例。

定义一个 MyActionFilter 如下：

```
public class MyActionFilter : IActionFilter
{
    public  void OnActionExecuting(ActionExecutingContext context)
{
        Debug.WriteLine("MyActionFilter.OnActionExecuting===>" );
```

```
    }
    public void OnActionExecuted(ActionExecutedContext context)
{
        Debug.WriteLine("MyActionFilter.OnActionExecuted===>" );
    }
}
```

只需要继承 IactionFilter 的接口并实现它的 OnActionExecuting 和 OnActionExecuted 两个方法即可。在各种 Filter 中，存在许多类似 On…Executing 和 On…Executed 的方法，一般前者表示在其负责的过程之前执行，后者在其之后执行。Action 的执行是被放在 ActionFilter 的这两个方法之间的。

为了验证这个例子，在默认的 Home.Index 这个 Action 中添加了 Debug 输出，代码如下：

```
public IActionResult Index()
{
    Debug.WriteLine("Executing===>Home.Index");
    return View();
}
```

Filter 定义之后需要将其注册到系统中，否则它不会生效。在 Startup 的 ConfigureServices 方法中做如下修改：

```
public void ConfigureServices(IServiceCollection services)
{
    services.AddControllersWithViews(options=> {
        options.Filters.Add<MyActionFilter>();
    });
}
```

访问任意一个 Home.Index，可以看到类似如下的内容：

```
MyActionFilter.OnActionExecuting===>
Executing===>Home.Index
MyActionFilter.OnActionExecuted====>
```

从输出结果中可知，在Action被执行的前后执行了OnActionExecuting和OnActionExecuted 两个方法。也可以用其他任意一个 Action 做验证，都会得到类似的输出结果，说明 MyActionFilter 已经被注册成“全局”的，对任何一个 Action 的访问都生效。

下面来看 Filter 的“复杂”用法。

19.3 Filter 的用法详解

在上面的例子中，我们注册了一个“全局”的 Filter，对任意一个 Action 的访问都生效。这里有一个问题，当多次访问时，生效的 Filter 是同一个实例还是多个不同的实例呢？

19.3.1 单例验证

为了验证这个问题，修改上文定义的 Filter，新的代码如下：

```
public class MyActionFilter : IActionFilter
{
    public Guid guid = Guid.NewGuid();

    public  void OnActionExecuting(ActionExecutingContext context)
    {
        // 省略处理代码
        Debug.WriteLine("MyActionFilter.OnActionExecuting===>"  + guid.ToString());
    }

    public void OnActionExecuted(ActionExecutedContext context)
    {
        // 省略处理代码
        Debug.WriteLine("MyActionFilter.OnActionExecuted====>"  + guid.ToString());
    }
}
```

添加了一个属性“[Guid guid]”，通过输出其值来验证是否是同一个实例。

多次访问多个 Action，都可以看到类似如下的内容，只是 Guid 的值不同。

```
MyActionFilter.OnActionExecuting===>29761b04-6560-4355-ae6c-17b383e0460c
Executing===>Home.Index MyActionFilter.OnActionExecuted====>29761b04-6560-4355-ae6c-
17b383e0460c
```

说明每次请求都产生了一个新的实例被执行。

从注册方式上看，我们将其作为一个“类型”注册，联想依赖注入，是不是可以将其作为实例注册呢？再次修改 AddMvc 方法：

```
services. AddControllersWithViews (option=>option.Filters.Add(new MyActionFilter()))
```

这里通过 new MyActionFilter()的方式创建了一个实例进行注册。经过多次任意访问、多个 Action 进行测试，发现 Guid 的值始终是同一个，说明这种实例方式注册后是“单例”的。

19.3.2 通过 Attribute 方式定义与注册

如何将 Filter 的作用域变为“非全局”的？也就是如何使 Filter 只针对特定的 Controller 或 Action 生效，而不对所有的 Action 都生效。可以通过 Attribute 的方式来实现，复制并修改上例的代码，生成一个新的 MyActionFilterAttribute，注意这里要使其继承 Attribute：

```
public class MyActionFilterAttribute : Attribute, IActionFilter
{
    public Guid guid = Guid.NewGuid();
```

```
    public void OnActionExecuting(ActionExecutingContext context)
    {
        Debug.WriteLine("MyActionFilter.OnActionExecuting===>" + guid.ToString());
    }

    public void OnActionExecuted(ActionExecutedContext context)
    {
        Debug.WriteLine("MyActionFilter.OnActionExecuted====>" + guid.ToString());
    }
}
```

这样就可以将其作为 Attribute 设置到 Controller 或 Action 上了。例如：

```
public class BookController : Controller
{
    [MyActionFilter]
    public string GetName(string code)
    {
        return "Net Core";
    }
}
```

不需要将其注册至 Startup，直接访问 Action，可以看到如下输出：

```
MyActionFilter.OnActionExecuting===>a334672d-c738-46ef-bc24-a9bf3be6de85
MyActionFilterAttribute.OnActionExecuting===>9de43951-b450-4c16-805f-4852068aa4c0
Executing===>Home.Index
MyActionFilterAttribute.OnActionExecuted====>9de43951-b450-4c16-805f-4852068aa4c0
MyActionFilter.OnActionExecuted====>a334672d-c738-46ef-bc24-a9bf3be6de85
```

而访问其他 Action，不会看到 MyActionFilterAttribute 的输出。若将其注册至 Controller，例如：

```
[MyActionFilter]
public class BookController : Controller
```

这样设置可以使 MyActionFilterAttribute 对此 Controller 的所有 Action 生效。若 Action 中已经设置，则 MyActionFilterAttribute 会被执行两次（这样的情况基本不会出现）。

19.3.3　支持继承方式注册

对于多个 Controller 都需要使用 MyActionFilterAttribute 的情况，我们当然不想对这些 Controller 逐个进行设置，一个简单的方式是采用继承来实现。首先定义一个 MyBaseController，并对其设置 MyActionFilterAttribute：

```
[MyActionFilter]
public class MyBaseController : Controller
```

然后让所有需要采用 MyActionFilterAttribute 的 Controller 都继承 MyBaseController，例如上文的 BookController，这些 Controller 就不需要逐一设置[MyActionFilter]了。例如：

```
public class BookController: MyBaseController
```

有人说这样仍需要对每个 Controller 设置继承，一样麻烦。对于当前情况来看，确实如此，貌似“工作量”是一样的。但在实际项目中，我们经常需要对 MyBaseController 这个“基类”进行更多的设置与操作，这时就体现出这样设计的优势了。

19.3.4 多功能 Filter

既然 Filter 的定义是实现 *IXXX*Filter 接口，那么是否可以用一个类来实现多个类型的接口呢？答案是可以的，比如在上例的基础上，再添加对 IResultFilter 的实现，生成一个新的 MyActionAndResultFilterAttribute，其代码如下：

```
public class MyActionAndResultFilterAttribute : Attribute, IActionFilter,IResultFilter
{
    public Guid guid = Guid.NewGuid();

    public void OnActionExecuting(ActionExecutingContext context)
    {
        Debug.WriteLine("MyActionAndResultFilterAttribute.OnActionExecuting===>");
    }

    public void OnActionExecuted(ActionExecutedContext context)
    {
        Debug.WriteLine("MyActionAndResultFilterAttribute.OnActionExecuted====>");
    }

    public void OnResultExecuting(ResultExecutingContext context)
    {
        Debug.WriteLine("MyActionAndResultFilterAttribute.OnResultExecuting====>");
    }

    public void OnResultExecuted(ResultExecutedContext context)
    {
        Debug.WriteLine("MyActionAndResultFilterAttribute.OnResultExecuted====>");
    }
}
```

增加了实现 IResultFilter 对应的 OnResultExecuting 和 OnResultExecuted 接口。将其添加到 Action 上进行验证：

```
public class BookController : Controller
{
```

```
    [MyActionAndResultFilter]
    public string GetName(string code)
    {
        return "Net Core";
    }
}
```

访问这个 GetName 的 Action，输出结果如下：

```
MyActionAndResultFilterAttribute.OnActionExecuting===>
Executing===>Home.Index
MyActionAndResultFilterAttribute.OnActionExecuted====>
MyActionAndResultFilterAttribute. OnResultExecuting ====>
。。。。。。 Executing ObjectResult, writing value of type 'System.String'.
MyActionAndResultFilterAttribute. OnResultExecuted ====>
```

19.3.5 Filter 的同步与异步

在上面的例子中，我们实现了 IactionFilter。Filter 都是成对出现的，一个同步，一个异步，比如对应 IActionFilter，还有一个异步的接口 IAsyncActionFilter。实现的方法也是异步的，例如 IAsyncActionFilter 接口如下：

```
public interface IAsyncActionFilter : IFilterMetadata
{
    Task OnActionExecutionAsync(ActionExecutingContext context, ActionExecutionDelegate
next);
}
```

只有一个名为 OnActionExecutionAsync 的异步方法，同样可以通过如下方式实现它：

```
public class MyAsyncActionFilter :Attribute , IAsyncActionFilter
{
    public Guid guid = Guid.NewGuid();

    public async Task OnActionExecutionAsync(ActionExecutingContext context,
ActionExecutionDelegate next)
    {
        Debug.WriteLine("MyActionFilter.OnActionExecutionAsync====> " );

        await next();
    }
}
```

从语法上来说，一个 Filter 可以同时实现一对同步和异步方法。例如下面的代码：

```
public class MyAsyncActionFilter :Attribute ,IActionFilter, IAsyncActionFilter
```

但在实际使用过程中，如果一个 Filter 同时实现了同一种 Filter 的同步和异步接口，在系统执行时会首先判断是否存在异步方法，若存在，则执行异步方法，否则才会执行同步方法。也就是说二者不会都被执行，具体原因在下文的代码分析中会说明。

19.3.6 继承内置 FilterAttribute

为了更方便地自定义 FilterAttribute，系统提供了一些内置的“基类”FilterAttribute。当需要自定义时，只需要继承对应的“基类”并重写对应的方法即可。它们是 ActionFilterAttribute、ExceptionFilterAttribute、ResultFilterAttribute、FormatFilterAttribute、ServiceFilterAttribute、TypeFilterAttribute。下面依然以 ActionFilterAttribute 为例，其代码如下：

```
[AttributeUsage(AttributeTargets.Class | AttributeTargets.Method, AllowMultiple = true,
Inherited = true)]
public abstract class ActionFilterAttribute :  Attribute, IActionFilter,
IAsyncActionFilter, IResultFilter, IAsyncResultFilter, IOrderedFilter
{
    /// <inheritdoc />
    public int Order { get; set; }

    /// <inheritdoc />
    public virtual void OnActionExecuting(ActionExecutingContext context)
    {
    }

    /// <inheritdoc />
    public virtual void OnActionExecuted(ActionExecutedContext context)
    {
    }

    /// <inheritdoc />
    public virtual async Task OnActionExecutionAsync(
        ActionExecutingContext context,
        ActionExecutionDelegate next)
    {
        if (context == null)
        {
            throw new ArgumentNullException(nameof(context));
        }

        if (next == null)
        {
            throw new ArgumentNullException(nameof(next));
        }

        OnActionExecuting(context);
```

```
        if (context.Result == null)
        {
            OnActionExecuted(await next());
        }
    }

    /// <inheritdoc />
    public virtual void OnResultExecuting(ResultExecutingContext context)
    {
    }

    /// <inheritdoc />
    public virtual void OnResultExecuted(ResultExecutedContext context)
    {
    }

    /// <inheritdoc />
public virtual async Task OnResultExecutionAsync(
        ResultExecutingContext context,
        ResultExecutionDelegate next)
{
        if (context == null)
        {
            throw new ArgumentNullException(nameof(context));
        }

        if (next == null)
        {
            throw new ArgumentNullException(nameof(next));
        }

        OnResultExecuting(context);
        if (!context.Cancel)
        {
            OnResultExecuted(await next());
        }
    }
}
```

可以看出，这个内置的 ActionFilterAttribute 主要继承并实现了 IactionFilter、IAsyncActionFilter，以及 IResultFilter、IAsyncResultFilter 这两对接口。显然带“Async”的是两个对应的异步 Filter 接口，只有两个异步方法内有内容，其作用仅是调用对应的同步方法。下面举例说明为什么要这样写：

```
public class NewActionFilter : ActionFilterAttribute
{
    public override void OnActionExecuting(ActionExecutingContext context)
```

```
        {
            Debug.WriteLine("NewActionFilter===========> OnActionExecuting ";
        }
        public override void OnActionExecuted(ActionExecutedContext context)
        {
            Debug.WriteLine("NewActionFilter===========>OnActionExecuted");
        }

        public override void OnResultExecuting(ResultExecutingContext context)
        {
            Debug.WriteLine("NewActionFilter===========>OnResultExecuting");
        }

        public override void OnResultExecuted(ResultExecutedContext context)
        {
            Debug.WriteLine("NewActionFilter===========>OnResultExecuted");
        }
    }
```

以上代码重写了 IActionFilter 和 IResultFilter 两个同步接口，同时实现了同步和异步两个方法，同步接口是不会被调用的。虽然我们自定义的 NewActionFilter 没有实现异步方法，但由于其“基类”ActionFilterAttribute 实现了异步方法，系统还是不会主动调用同步方法。所以 ActionFilterAttribute 的异步方法中写了对同步方法的调用，这样我们通过继承 ActionFilterAttribute 来自定义 Filter 时，就可以自由选择是实现同步方法还是异步方法了。

ASP.NET Core 提供的其他内置 FilterAttribute 如下。

- ActionFilterAttribute
- ExceptionFilterAttribute
- ResultFilterAttribute
- FormatFilterAttribute
- ServiceFilterAttribute
- TypeFilterAttribute

用法同理，这里就不一一介绍了。

19.4 Filter 的获取

下面通过示例来讲解 Filter 被注册之后是如何被系统获取并执行的。为了更明显地看出区别，示例代码尽量写得简单，如下：

```
    public class DemoActionFilter1 : Attribute, IActionFilter
    {
        public void OnActionExecuting(ActionExecutingContext context)
        {
```

```
            Debug.WriteLine("DemoActionFilter1===>OnActionExecuting");
        }
        public void OnActionExecuted(ActionExecutedContext context)
        {
            Debug.WriteLine("DemoActionFilter1===>OnActionExecuted");
        }
    }
    public class DemoActionFilter2 : Attribute, IActionFilter
    {
        public void OnActionExecuting(ActionExecutingContext context)
        {
            Debug.WriteLine("DemoActionFilter2===>OnActionExecuting");
        }
        public void OnActionExecuted(ActionExecutedContext context)
        {
            Debug.WriteLine("DemoActionFilter2===>OnActionExecuted");
        }
    }
    // 省略 DemoActionFilter3
    public class DemoResultFilter1 : Attribute, IResultFilter
    {
        public void OnResultExecuting(ResultExecutingContext context)
        {
            Debug.WriteLine("DemoResultFilter1===>OnResultExecuting");
        }

        public void OnResultExecuted(ResultExecutedContext context)
        {
            Debug.WriteLine("DemoResultFilter1===>OnResultExecuted");
        }
    }
      // 省略 DemoResultFilter2

      // 省略 DemoResultFilter3
```

共创建了 3 个 ActionFilter 和 3 个 ResultFilter。现将它们进行注册，首先在 Startup 中注册 DemoActionFilter1 和 DemoResultFilter1：

```
services. AddControllersWithViews (
    options => {
        options.Filters.Add(new DemoActionFilter1());
        options.Filters.Add(new DemoResultFilter1());
    }
)
```

然后在 Controller 中注册：

```
[DemoActionFilter2]
[DemoResultFilter2]
public class BookController : Controller
{
    [DemoResultFilter3]
    [DemoActionFilter3]
    public string GetName(string code)
    {
        return "Net Core";
    }
}
```

按照 Startup、Controller、Action 注册的顺序，将两种 Filter 按照编号 1、2、3 的顺序注册，然后开始访问 GetName 的 Action，看这几个 Filter 是在什么时候被获取的。

在第 17 章 Action 的执行中，有一个过程是 invoker 的生成，如图 17-1 所示。invoker 有一个重要的属性 filters，即被请求的 Action 对应的所有 Filter 的集合。获取 filters 的代码如下（在 ControllerActionInvokerCache.cs 中）：

```
public (ControllerActionInvokerCacheEntry cacheEntry, IFilterMetadata[] filters)
GetCachedResult(ControllerContext controllerContext)
        {
            var cache = CurrentCache;
            var actionDescriptor = controllerContext.ActionDescriptor;
            IFilterMetadata[] filters;
            if (!cache.Entries.TryGetValue(actionDescriptor, out var cacheEntry))
            {
                var filterFactoryResult = FilterFactory.GetAllFilters(_filterProviders,
controllerContext);
                filters = filterFactoryResult.Filters;
                //省略关于其他属性值的获取
}
}
```

FilterFactory 的 GetAllFilters 方法（FilterFactory.cs）如下：

```
 public static FilterFactoryResult GetAllFilters(IFilterProvider[] filterProviders,
 ActionContext actionContext)
 {
     //省略
     var actionDescriptor = actionContext.ActionDescriptor;
     var orderedFilters = actionDescriptor.FilterDescriptors.OrderBy(filter => filter,
FilterDescriptorOrderComparer.Comparer).ToList();
     //省略
     return new FilterFactoryResult(staticFilterItems, filters);
 }
```

这里保留了关键的几行代码，也就是根据被请求的 Action 的 ActionDescriptor 来获取它对应的所有 Filter（无论是在 Startup、Controller 还是在 Action 中注册的）。对这些 Filter 进行排序，这里用到了排序方法 FilterDescriptorOrderComparer，它用来定义 Filter 的执行顺序，19.5 节会详细介绍。

本例的情况会获取 9 个 Filter，如图 19-2 所示。

```
[0] {Microsoft.AspNetCore.Mvc.Internal.ControllerActionFilter}
[1] {Microsoft.AspNetCore.Mvc.ViewFeatures.Internal.SaveTempDataFilter}
[2] {Microsoft.AspNetCore.Mvc.ModelBinding.UnsupportedContentTypeFilter}
[3] {FilterDemo.DemoActionFilter1}
[4] {FilterDemo.DemoResultFilter1}
[5] {FilterDemo.DemoActionFilter2}
[6] {FilterDemo.DemoResultFilter2}
[7] {FilterDemo.DemoResultFilter3}
[8] {FilterDemo.DemoActionFilter3}
```

▲图 19-2

系统自动添加了前 3 个 Filter，后 6 个是自定义注册的，注意这已经是排序后的结果，读者可以自行调整注册顺序进行验证。系统自动添加的 3 个 Filter 如下。

- **ControllerActionFilter**：用于调用 Controller 的 ActionFilter 接口。
- **SaveTempDataFilter**：用于 TempData 的处理。
- **UnsupportedContentTypeFilter**：用于不支持的参数类型处理，主要是对应 UnsupportedContentTypeException 异常。

invoker 获取了当前请求对应的所有 Filter 并进行排序，获取阶段结束。

19.5 Filter 的执行

Filter 的执行在 invoker 的执行阶段，见 17.3 节的 invoker 的执行。请求会进入 FilterPipeline（InvokeFilterPipelineAsync 方法）进行处理。FilterPipeline 被称为“筛选器管道”，作用是通过一定的顺序执行这些 Filter 以及 Action，结构如图 19-3 所示。

顺序的设定通过 ResourceInvoker 和 ControllerActionInvoker 中的两个 while 循环分别调用两个复杂的 Next 方法实现。由于这几段代码很长，将代码精简后来看其结构。

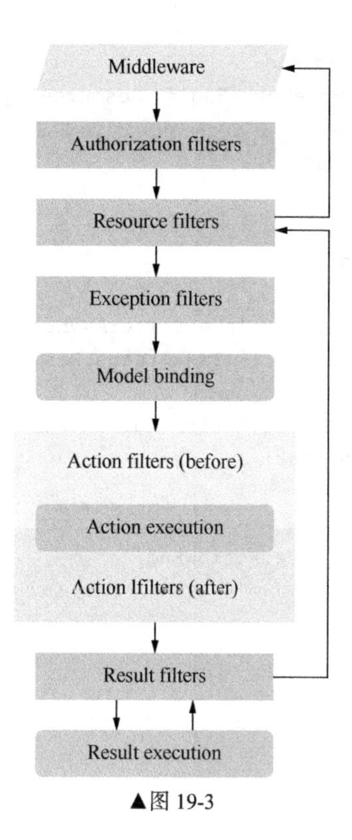

▲图 19-3

ResourceInvoker 的 Next 方法如下：

```
private Task Next(ref State next, ref Scope scope, ref object state, ref bool isCompleted)
{
    switch (next)
    {
        case State.InvokeBegin:

        case State.AuthorizationBegin:

        case State.AuthorizationNext:

        case State.AuthorizationAsyncBegin:

        case State.AuthorizationAsyncEnd:

        case State.AuthorizationSync:

        case State.AuthorizationShortCircuit:

        case State.AuthorizationEnd:
```

```
        case State.ResourceBegin:
        //ResourceNext、ResourceAsyncBegin、ResourceAsyncEnd……

        case State.ExceptionBegin:
        //Exception……

        case State.ActionBegin:
            {
                var task = InvokeInnerFilterAsync();
                if (task.Status != TaskStatus.RanToCompletion)
                {
                    next = State.ActionEnd;
                    return task;
                }

                goto case State.ActionEnd;
            }

        case State.ActionEnd:

        case State.ResourceInsideEnd:

        case State.ResourceEnd:

        case State.InvokeEnd:

    }
}
```

ControllerActionInvoker 的 Next 方法如下：

```
private Task Next(ref State next, ref Scope scope, ref object state, ref bool isCompleted)
{
    switch (next)
    {
        case State.ActionBegin:

        case State.ActionNext:

        case State.ActionAsyncBegin:

        case State.ActionAsyncEnd:

        case State.ActionSyncBegin:

        case State.ActionSyncEnd:
```

```
        case State.ActionInside:

        case State.ActionEnd:

        default:
            throw new InvalidOperationException();
    }
}
```

通过上面的几段代码可以看出，ResourceInvoker 的 Next 方法用于 Authorization、Resource、Exception 等类型的 Filter，而 ControllerActionInvoker 的 Next 方法主要涉及 Action Filters。在 ResourceInvoker 的 Next 方法中的 case State.ActionBegin 是一个外壳，用于调用 ControllerAction Invoker 的 InvokeInnerFilterAsync 方法，也就进入了 ControllerActionInvoker 的 Next 方法。

从两个 Next 方法的各个“case”中可以发现，每种类型都对应着类似 *XXX*Begin、*XXX*Next、*XXX*AsyncBegin、*XXX*AsyncEnd 等 case。虽然每种类型的实际应用方式不同，它们的代码细节上有一些不同，但大体运行思路类似。

本节依然以 Action Filters 为例来讲解，也就是 ControllerActionInvoker 的 Next 方法。对于其他类型的 Filter 的处理，读者可以自行类比研究。

首先，从“case State.ActionBegin”开始，一般 *XXX*Begin 用于做一些准备工作，代码如下：

```
case State.ActionBegin:
    {
        var controllerContext = _controllerContext;

        _cursor.Reset();

        _instance = _cacheEntry.ControllerFactory(controllerContext);

        _arguments = new Dictionary<string, object>(StringComparer.OrdinalIgnoreCase);

        var task = BindArgumentsAsync();
        if (task.Status != TaskStatus.RanToCompletion)
        {
            next = State.ActionNext;
            return task;
        }

        goto case State.ActionNext;
    }
```

controllerContext 是对请求上下文 httpContext、目标 action 的信息 ActionDescriptor，以及路由信息的封装。_cacheEntry.ControllerFactory 是 invoker 生成阶段创建的一个用于创建 Controller 的方法，所以 instance 就是当前请求的目标 Controller。最后调用 BindArgumentsAsync()

进行参数绑定。

其次，查找需要执行的 IActionFilter 或 IasyncActionFilter：

```
case State.ActionNext:
    {
        var current = _cursor.GetNextFilter<IActionFilter, IAsyncActionFilter>();
        if (current.FilterAsync != null)
        {
            if (_actionExecutingContext == null)
            {
                _actionExecutingContext = new ActionExecutingContext(_controllerContext,
_filters, _arguments, _instance);
            }

            state = current.FilterAsync;
            goto case State.ActionAsyncBegin;
        }
        else if (current.Filter != null)
        {
            if (_actionExecutingContext == null)
            {
                _actionExecutingContext = new ActionExecutingContext(_controllerContext,
_filters, _arguments, _instance);
            }

            state = current.Filter;
            goto case State.ActionSyncBegin;
        }
        else
        {
            goto case State.ActionInside;
        }
    }
```

调用_cursor 的 GetNextFilter<IActionFilter，IAsyncActionFilter>方法，这是本阶段的核心。在“State.ActionBegin”阶段就出现过_cursor，调用了它的 Reset 方法。_cursor 是一个用于在 Filter 的集合中查找指定类型的 Filter 的游标，这个集合就是 19.4 节中已经获取并排序的 Filter 的集合。_cursor 的代码类型为 FilterCursor：

```
public struct FilterCursor
    {
        private readonly IFilterMetadata[] _filters;
        private int _index;
        public FilterCursor(IFilterMetadata[] filters)
        {
            _filters = filters;
```

```
            _index = 0;
        }
        public void Reset()
        {
            _index = 0;
        }

        public FilterCursorItem<TFilter, TFilterAsync> GetNextFilter<TFilter,
TFilterAsync>()
            where TFilter : class
            where TFilterAsync : class
        {
            while (_index < _filters.Length)
            {
                var filter = _filters[_index] as TFilter;
                var filterAsync = _filters[_index] as TFilterAsync;

                _index += 1;

                if (filter != null || filterAsync != null)
                {
                    return new FilterCursorItem<TFilter, TFilterAsync>(filter,
filterAsync);
                }
            }

            return default(FilterCursorItem<TFilter, TFilterAsync>);
        }
    }
```

根据以上代码，_cursor 的处理逻辑如图 19-4 所示。

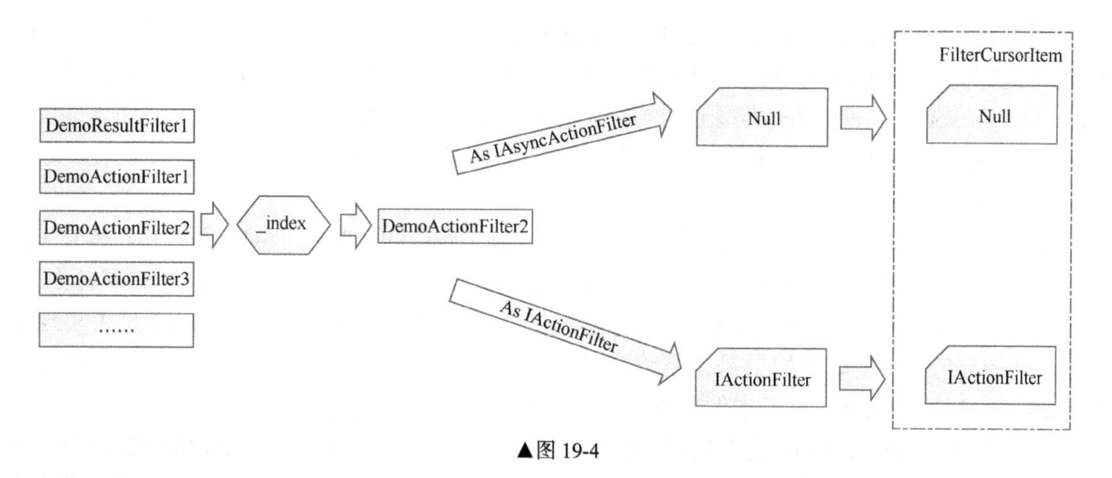

▲图 19-4

FilterCursor 有一个内置的 int _index 变量，用于记录当前已经查找过的位置，在 GetNextFilter 方法中，通过这个变量对 Filter 的集合进行遍历。在图 19-4 中，当获取的结果为 DemoActionFilter2 时，会将 DemoActionFilter2 分别转换为 IActionFilter 和 IAsyncActionFilter，即 ActionFilter 的同步和异步接口类型。因为 ActionFilter 只实现了 IActionFilter 接口，所以转换为 IActionFilter 后的结果不为空，而另一个转换为 IAsyncActionFilter 的结果为空。若二者至少有一个结果不为空，则将其封装为 FilterCursorItem 返回；若均为空，即未实现任何一个接口，则继续判断下一个是否是想要的类型。例如当获取的结果为 DemoResultFilter1 时，是不符合要求的，会跳过并判断下一个。

FilterCursorItem 的代码如下，其中有两个属性分别对应参数中的同步和异步接口，被转换后的两个结果分别赋值给对应的属性。所以图 19-4 中 DemoActionFilter2 被封装为 FilterCursorItem 后，它的 TFilter Filter 属性是有值的，而 TFilterAsync FilterAsync 属性为空。若遍历到 Filter 的集合的结尾依然没有找到符合条件的，会返回一个默认的 FilterCursorItem，它的两个对应属性值都为空。

```
public readonly struct FilterCursorItem<TFilter, TFilterAsync>
{
    public FilterCursorItem(TFilter filter, TFilterAsync filterAsync)
    {
        Filter = filter;
        FilterAsync = filterAsync;
    }

    public TFilter Filter { get; }

    public TFilterAsync FilterAsync { get; }
}
```

不只在处理 ActionFilter 时，在处理 Authorization、Resource、Exception 等类型的 Filter 时也是这样的。通过 FilterCursor 的 GetNextFilter<TFilter，TFilterAsync>方法的参数也可以看出，FilterCursor 是多种 Filter 类型通用的，所以在准备处理一种新的 Filter 类型时会把它重置，也就是调用_cursor.Reset()，将_cursor 的_index 属性值重置为 0。这一般会发生在“case State.*XXX*Begin”阶段。

下面来看“case State.ActionNext”的代码。当找到了对应的 Filter，例如在图 19-4 得到了由 DemoActionFilter2 封装的 FilterCursorItem 后，将其赋值给 current。接下来是一个 if 语句，具体如下。

（1）判断 current 的 FilterAsync 属性是否为空，也就是判断 DemoActionFilter2 是否实现了 IAsyncActionFilter 接口，如果是，则将它的 FilterAsync 属性赋值给 State，并跳转到 case State.ActionAsyncBegin。对于本例来说，这个结果是 false，会进行后面的判断。

（2）接下来判断 current 的 Filter 属性是否为空，也就是判断 DemoActionFilter2 是否实现了 IActionFilter 接口，如果是，则将它的 FilterAsync 属性（也就是 DemoActionFilter2）赋值给

State，并跳转到 case State.ActionSyncBegin。对于本例来说，这个结果是 true，接下来流程跳转到 case State.ActionSyncBegin。

（3）若上面的两条结果均为 false，则说明在 Filter 的集合中，_index 标识的位置之后没有符合作用条件的 Filter，则会跳转至 case State.ActionInside，也就是执行 Action 方法。

这样 if 语句的作用是判断是否还有满足条件的 Filter，若有，则判断其是否实现了异步或同步接口。若实现，则进入对应的异步或同步的 case 阶段。若没有满足条件，则进入 Action 的方法调用阶段 case State.ActionInside。无须同时实现 Filter 的同步和异步接口，是因为会首先判断是否实现了异步接口，未实现才会调用同步接口，原因就在于 if 条件的设置。

继续进行 DemoActionFilter2 的例子，由于其实现了同步接口，流程进入 case State.ActionSyncBegin 阶段。删除了日志相关代码后如下：

```
case State.ActionSyncBegin:
    {
        var filter = (IActionFilter)state;
        var actionExecutingContext = _actionExecutingContext;
        filter.OnActionExecuting(actionExecutingContext);
        if (actionExecutingContext.Result != null)
        {
            _actionExecutedContext = new ActionExecutedContext(
                _actionExecutingContext,
                _filters,
                _instance)
            {
                Canceled = true,
                Result = _actionExecutingContext.Result,
            };

            goto case State.ActionEnd;
        }

        var task = InvokeNextActionFilterAsync();
        if (task.Status != TaskStatus.RanToCompletion)
        {
            next = State.ActionSyncEnd;
            return task;
        }

        goto case State.ActionSyncEnd;
    }
```

首先获取上一阶段保存的 State，也就是 DemoActionFilter2。接下来调用 DemoActionFilter2 的 OnActionExecuting 方法，参数为 actionExecutingContext。执行后判断 actionExecutingContext.Result 是否不为空，即是否已经给出了请求的响应结果。若是，则进入 case State.ActionEnd，这个过程相

当于对请求流程进行了短路；否则会调用 InvokeNextActionFilterAsync 方法获取下一个 Filter。之后进入 case State.ActionSyncEnd 阶段：

```
case State.ActionSyncEnd:
    {
        var filter = (IActionFilter)state;
        var actionExecutedContext = _actionExecutedContext;
        filter.OnActionExecuted(actionExecutedContext);
        goto case State.ActionEnd;
    }
```

这个阶段的主要作用是调用 Filter 的 OnActionExecuted 方法。

InvokeNextActionFilterAsync 方法的代码如下：

```
private async Task InvokeNextActionFilterAsync()
{
    var next = State.ActionNext;
    var state = (object)null;
    var scope = Scope.Action;
    var isCompleted = false;
    while (!isCompleted)
    {
        await Next(ref next, ref scope, ref state, ref isCompleted);
    }
}
```

它的作用是将流程重新指向 case State.ActionNext，也就是 Filter 的获取阶段。由于游标不会被重置，所以会继续寻找 DemoActionFilter2 后面的实现了 IActionFilter 或 IAsyncActionFilter 接口的 Filter。如图 19-4 所示，下一个应该是 DemoActionFilter3。执行完 DemoActionFilter3 后，再次进入同样的流程循环，直至找不到这样的 Action Filter 为止。这时按照 case State.ActionNext 阶段的 if 语句的第 3 条，即最终找不到符合条件的 Filter 时，就会进入 case State.ActionInside 阶段，调用 InvokeActionMethodAsync 方法来执行 Action 中的内容。代码如下：

```
case State.ActionInside:
    {
        var task = InvokeActionMethodAsync();
        if (task.Status != TaskStatus.RanToCompletion)
        {
            next = State.ActionEnd;
            return task;
        }

        goto case State.ActionEnd;
    }
```

这样的嵌套调用形成了与中间件组成的请求处理管道一样的结构，如图 19-5 所示。

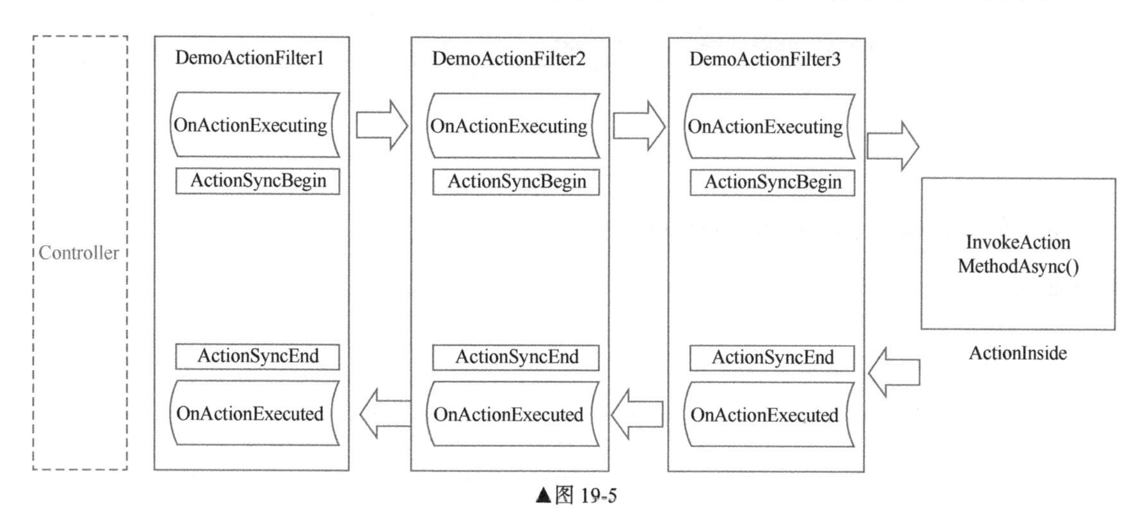

▲图 19-5

至此，Filter 的执行基本上讲完了。按照图 19-5 所示的顺序，在 Action 的执行前后分别执行了 Action Filter 的 OnActionExecuting 和 OnActionExecuted 方法。这里没有讲异步的 IAsyncActionFilter 接口的 Filter 的执行，它会进入对应的 case State.ActionAsyncBegin 和 case State.ActionAsyncEnd 阶段，代码和同步的两个对应阶段类似。

注意图 19-5 的最左边有一个虚线标注的 Controller，因为 Controller 实现了 Action 的 Filter 接口，Controller 和 ActionFilterAttribute 一样同时实现了同步和异步接口，并在异步接口中添加了对同步接口的调用，方便我们重写这几个方法。标注为虚线是因为当 BookController 只作为 WebAPI 时（派生自 ControllerBase 而不是 Controller），此处不会出现 Controller。精简后的 Controller 代码如下：

```
public abstract class Controller : ControllerBase, IActionFilter, IAsyncActionFilter,
 IDisposable
    {
       public virtual void OnActionExecuting(ActionExecutingContext context)
       {
       }
       public virtual void OnActionExecuted(ActionExecutedContext context)
       {
       }
       public virtual async Task OnActionExecutionAsync(
           ActionExecutingContext context,
           ActionExecutionDelegate next)
       {
           //省略验证代码

           OnActionExecuting(context);
```

```
            if (context.Result == null)
            {
                OnActionExecuted(await next());
            }
        }

    }
```

可以在 BookController 中重写对应的方法。

```
public class BookController : Controller
    {
    //省略其他 Action 及方法

        public override void OnActionExecuting(ActionExecutingContext context)
        {
            Debug.WriteLine("BookController=======>OnActionExecuting" );
        }
        public override void OnActionExecuted(ActionExecutedContext context)
        {
            Debug.WriteLine("BookController=======>OnActionExecuted");
        }
}
```

Controller 之所以会在图 19-5 的最左边（也就是最先执行），是因为获取的 Filter 的集合的第一项是 ControllerActionFilter，即它是被系统默认添加的一个 Filter，代码如下：

```
public class ControllerActionFilter : IAsyncActionFilter, IOrderedFilter
{
    public int Order { get; set; } = int.MinValue;

    public Task OnActionExecutionAsync(
        ActionExecutingContext context,
        ActionExecutionDelegate next)
    {
        if (context == null)
        {
            throw new ArgumentNullException(nameof(context));
        }

        if (next == null)
        {
            throw new ArgumentNullException(nameof(next));
        }

        var controller = context.Controller;
        if (controller == null)
```

```
        {
            throw new InvalidOperationException(Resources.FormatPropertyOfTypeCannotBeNull(
                nameof(context.Controller),nameof(ActionExecutingContext)));
        }

        if (controller is IAsyncActionFilter asyncActionFilter)
        {
            return asyncActionFilter.OnActionExecutionAsync(context, next);
        }
        else if (controller is IActionFilter actionFilter)
        {
            return ExecuteActionFilter(context, next, actionFilter);
        }
        else
        {
            return next();
        }
    }

    private static async Task ExecuteActionFilter(
        ActionExecutingContext context,
        ActionExecutionDelegate next,
        IActionFilter actionFilter)
    {
        actionFilter.OnActionExecuting(context);
        if (context.Result == null)
        {
            actionFilter.OnActionExecuted(await next());
        }
    }
}
```

ControllerActionFilter 实现了 IAsyncActionFilter 接口，并且排在第一个，所以它会被最先执行，即执行它的 OnActionExecutionAsync 方法。在这个方法里又会获取 Controller 并判断其是否实现了 IAsyncActionFilter 接口，若实现了，则调用 Controller 的 OnActionExecutionAsync 方法。

最后访问 Book/GetName，输出的结果如下：

```
BookController=======>OnActionExecuting
DemoActionFilter1===>OnActionExecuting
DemoActionFilter2===>OnActionExecuting
DemoActionFilter3===>OnActionExecuting
......Executing action method......BookController.GetName......
BookController=======>GetName //此处 Action 方法被执行
......Executed action method.......BookController.GetName......
DemoActionFilter3===>OnActionExecuted
```

```
DemoActionFilter2===>OnActionExecuted
DemoActionFilter1===>OnActionExecuted
BookController=======>OnActionExecuted

//上面是 ActionFilter 相关，下面是 ResultFilter 相关

DemoResultFilter1===>OnResultExecuting
DemoResultFilter2===>OnResultExecuting
DemoResultFilter3===>OnResultExecuting
……Executing ObjectResult, writing value of type 'System.String'.
DemoResultFilter3===>OnResultExecuted
DemoResultFilter2===>OnResultExecuted
DemoResultFilter1===>OnResultExecuted
```

输出结果验证了图 19-5 所示的 Filter 的执行顺序。

我们可以像 BookController 重写 Controller 的 OnActionExecuting 和 OnActionExecuted 方法一样，将一些通用的需要在 Action 执行前后执行的代码写在 BookController 的这两个方法里，而无须将其作为 Filter 注册。为了避免在每个 Controller 中添加这样的代码，可以新建一个 BaseController，将通用的需要执行的代码写在它的 OnActionExecuting 和 OnActionExecuted 方法里，然后所有具有此功能需要的 Controller 都继承 BaseController 即可。

19.6 Filter 的执行顺序

Filter 的执行顺序由 3 部分决定。

（1）对于不同类型的 Filter，按照图 19-3 所示的顺序执行，例如 Authorization filters 会最先被执行。

（2）对于相同类型的 Filter，执行顺序由其 Order 和 Scope 来决定。

Filter 的排序方法 FilterDescriptorOrderComparer 用于对获取的 Filter 的集合进行排序，相关代码如下：

```
    public class FilterDescriptorOrderComparer : IComparer<FilterDescriptor>
    {
        public static FilterDescriptorOrderComparer Comparer { get; } = new FilterDes
criptorOrderComparer();

        public int Compare(FilterDescriptor x, FilterDescriptor y)
        {
            if (x == null)
            {
                throw new ArgumentNullException(nameof(x));
            }

            if (y == null)
```

```
            {
                throw new ArgumentNullException(nameof(y));
            }

            if (x.Order == y.Order)
            {
                return x.Scope.CompareTo(y.Scope);
            }
            else
            {
                return x.Order.CompareTo(y.Order);
            }
        }
    }
```

从这个方法可以看出 Filter 的执行顺序，即按照先 Order、后 Scope 的方式排序。对于继承默认的内置 Filter 的，Order 默认为 0。默认情况下，全局的为 10、Controller 上的为 20、Action 上的为 30。也就是说，Filter 的执行顺序为“全局→Controller→Action”，所以它们的输出结果如下：

```
OnActionExecuting//全局

     Controller OnActionExecuting

          Action OnActionExecuting

          Action OnActionExecuted

     Controller OnActionExecuted

OnActionExecuted//全局
```

上面 DemoActionFilter1、DemoActionFilter2、DemoActionFilter3 的例子恰恰验证了这一点。

当然我们可以自定义 Filter 的 Order，使其不再采用默认值 0，这只需在其构造方法中设置，代码如下：

```
    public class Test1Filter : ActionFilterAttribute
    {
        public Test1Filter()
        {
            Order = 1;
        }

        // 省略部分代码

    }
```

（3）对于作用域和类型均一样的 Filter 来说，执行顺序是按照注册先后排列的。例如：

```
[Test2Filter]
[Test1Filter]
public JsonResult Index()
```

先执行 Test2Filter，后执行 Test1Filter。

第 20 章 控制返回类型

第 18 章介绍了系统如何将客户端提交的数据格式化处理成我们想要的类型，并绑定对应的参数。本章讲解它的“逆过程”，即如何将请求结果按照客户端想要的类型返回去。

20.1 常见的返回类型

以系统模板默认生成的 Home/Index 这个 Action 来说，为什么当请求它时会返回一个 HTML 页面呢？此外，还有 JSON、TXT 等类型，系统如何处理这些不同的类型呢？

20.1.1 返回类型

下面看几种常见的返回类型的例子，并用 Fiddler 请求观察结果，涉及一个名为 Book 的类，代码为：

```
public class Book
{
    public string Code { get; set; }
    public string Name { get; set; }
}
```

1. ViewResult 类型

ViewResult 类型代码为：

```
public class HomeController : Controller
{
    public IActionResult Index()
    {
        return View();
    }
}
```

返回一个 HTML 页面，Content-Type 值为：

```
text/html; charset=utf-8
```

2. string 类型

string 类型代码为：

```
public string GetString()
{
    return "Hello Core";
}
```

返回字符串“Hello Core”，Content-Type 值为：

```
text/plain; charset=utf-8
```

3. JSON 类型

JSON 类型代码为：

```
public JsonResult GetJson()
{
    return new JsonResult(new Book() { Code = "1001", Name = "ASP" });
}
```

返回 JSON，值为：

```
{"code":"1001","name":"ASP"}
```

Content-Type 值为：

```
application/json; charset=utf-8
```

4. 直接返回实体类型

JSON 类型代码为：

```
public Book GetModel()
{
    return new Book() { Code = "1001", Name = "ASP" };
}
```

同样返回 JSON，值为：

```
{"code":"1001","name":"ASP"}
```

Content-Type 值同样是：

```
application/json; charset=utf-8
```

5. void 类型

void 类型代码为：

```
public void GetVoid()
{
}
```

没有返回结果，也没有 Content-Type 值。

20.1.2　异步方法

1. Task 类型

Task 类型代码为：

```
public async Task GetTaskNoResult()
{
    await Task.Run(() => { });
}
```

与 void 类型一样，没有返回结果，也没有 Content-Type 值。

2. Task<string>类型

Task<string>类型代码为：

```
public async Task<string> GetTaskString()
{
    string rtn = string.Empty;

    await Task.Run(() => { rtn = "Hello Core"; });
    return rtn;
}
```

与 string 类型一样，返回字符串“Hello Core”，Content-Type 值为：

```
text/plain; charset=utf-8
```

3. Task<JsonResult>类型

Task<JsonResult>类型代码为：

```
public async Task<JsonResult> GetTaskJson()
{
    JsonResult jsonResult = null;
    await Task.Run(() => { jsonResult = new JsonResult(new Book() { Code = "1001",
Name = "ASP" }); });
    return jsonResult;
}
```

与 JSON 类型一样，返回 JSON，值为：

```
[{"code":"1001","name":"ASP"},{"code":"1002","name":"Net Core"}]
```

Content-Type 值为：

```
application/json; charset=utf-8
```

还有其他类型，总结如下。

- 返回结果有空、Html 页面、普通字符串、JSON 字符串等。
- 对应的 Content-Type 类型有空、text/html、text/plain、application/json 等。
- 异步 Action 的返回结果和其对应的同步 Action 返回结果类型一致。

20.2 内部处理机制解析

20.2.1 总体流程

下面来看系统是如何处理这些不同的类型的，图 20-1 所示为总体流程。

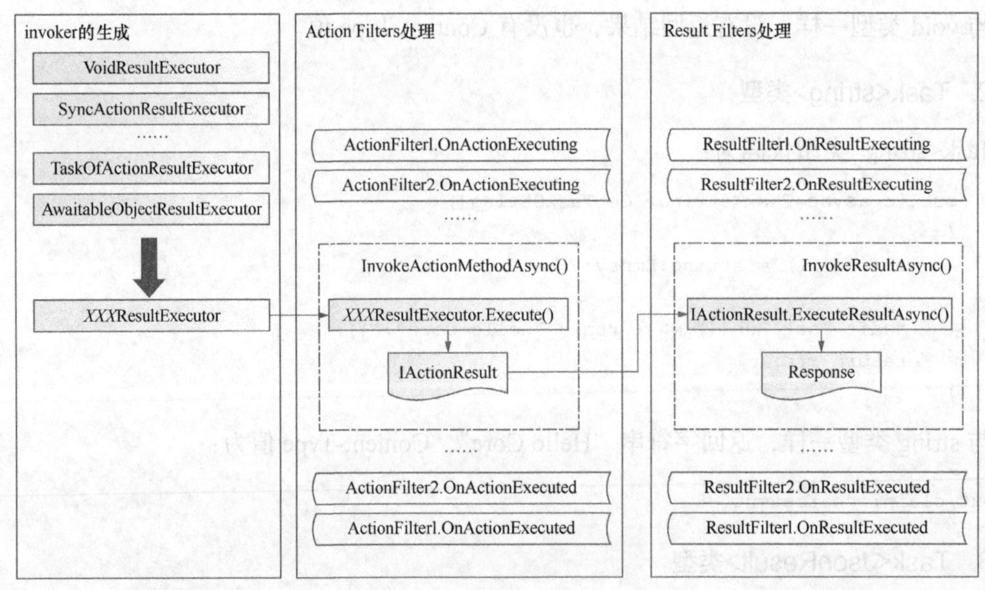

▲图 20-1

其中涉及 3 部分内容。

第一部分：invoker 的生成阶段。17.2 节中讲到了 Action 的执行者的获取，是从一系列系统定义的 *XXX*ResultExecutor 中筛选出来的（所在类为 ActionMethodExecutor）。虽然它们名为 *XXX*ResultExecutor，但都是 Action 的执行者，而不是 ActionResult 的执行者，都是 ActionMethodExecutor 的子类。以 Action 是同步、异步以及 Action 的返回值类型为筛选条件，具体见图 17-2 所示 *XXX*ResultExecutor 列表及其后面的筛选逻辑部分。在图 20-1 中，筛选出了被请求的 Action 对应的 *XXX*ResultExecutor，若以 Home/Index 这个默认的 Action 为例，*XXX*ResultExecutor 应该是 SyncActionResultExecutor。

第二部分：Action Filters 的处理阶段。这部分内容见 19.5 节，该节恰好以 Action Filter 为例讲了 Action Filter 的执行方式及 Action 被执行的过程。这个阶段会调用上文筛选出的 SyncActionResult Executor 的 Execute 方法来执行 Home/Index 这个 Action。执行结果返回 IActionResult。

第三部分：Result Filters 的处理阶段。这个阶段和 Action Filters 的逻辑相似，但前者的核心是 Action 的执行，后者的核心是 Action 的执行结果的执行。二者都分为 OnExecuting 和 OnExecuted 两个方法，且在其对应的核心执行方法前后执行。

20.2.2 ActionMethodExecutor 的选择与执行

第一部分中，系统为什么要定义这么多种 *XXX*ResultExecutor，并且在请求时逐个筛选合适的 *XXX*ResultExecutor 呢？从筛选规则是以 Action 的同步、异步，以及 Action 的返回值类型来看，这么多种 *XXX*ResultExecutor 是为了处理不同的 Action 类型。

依然以 Home/Index 为例，在筛选 *XXX*ResultExecutor 时，最终的返回结果是 SyncAction ResultExecutor。代码如下：

```
private class SyncActionResultExecutor : ActionMethodExecutor
        {
            public override ValueTask<IActionResult> Execute(
                IActionResultTypeMapper mapper,
                ObjectMethodExecutor executor,
                object controller,
                object[] arguments)
            {
                var actionResult = (IActionResult)executor.Execute(controller, arguments);
                EnsureActionResultNotNull(executor, actionResult);

                return new ValueTask<IActionResult>(actionResult);
            }

            protected override bool CanExecute(ObjectMethodExecutor executor)
                => !executor.IsMethodAsync && typeof(IActionResult).IsAssignableFrom
(executor.MethodReturnType);
        }
```

*XXX*ResultExecutor 的 CanExecute 方法是筛选的条件，通过这个方法判断它是否适合当前请求的目标 Action。它要求 Action 不是异步的，并且返回结果类型是派生自 IactionResult 的。而 Home/Index 这个 Action 标识的返回结果是 IActionResult，实际是通过 View 方法返回的，返回结果类型是 IActionResult 的派生类 ViewResult。该派生类还有常见的 JsonResult 和 ContentResult 等，它们都继承了 ActionResult，而 ActionResult 实现了 IActionResult 接口。所以如果一个 Action 是同步的，并且返回结果是 JsonResult 或 ContentResult 时，对应的 *XXX*Result

Executor 也是 SyncActionResultExecutor。

第二部分中，Action 的执行在 *XXX*ResultExecutor 的 Execute 方法中，它会进一步调用 ObjectMethodExecutor 的 Execute 方法。实际上所有的 Action 都是由 ObjectMethodExecutor 的 Execute 方法来执行的。而众多的 *XXX*ResultExecutor 方法的作用是调用这个方法，并对返回结果进行验证和处理，例如，SyncActionResultExecutor 会通过 EnsureActionResultNotNull 方法确保返回的结果不能为空。

string 类型对应的是 SyncObjectResultExecutor，代码如下：

```
private class SyncObjectResultExecutor : ActionMethodExecutor
{
    public override ValueTask<IActionResult> Execute(
        IActionResultTypeMapper mapper,
        ObjectMethodExecutor executor,
        object controller,
        object[] arguments)
    {
        var returnValue = executor.Execute(controller, arguments);
        var actionResult = ConvertToActionResult(mapper, returnValue,
executor.MethodReturnType);
        return new ValueTask<IActionResult>(actionResult);
    }

    protected override bool CanExecute(ObjectMethodExecutor executor)
=> !executor.IsMethodAsync;
}
```

由于 string 类型不是 IActionResult 的子类，所以会通过 ConvertToActionResult 方法对返回结果 returnValue 进行处理：

```
private IActionResult ConvertToActionResult(IActionResultTypeMapper mapper, object
returnValue, Type declaredType)
{
    var result = (returnValue as IActionResult) ?? mapper.Convert(returnValue,
declaredType);
    if (result == null)
    {
        throw new InvalidOperationException(Resources.FormatActionResult_ActionReturn
ValueCannotBeNull(declaredType));
    }

    return result;
}
```

如果 returnValue 是 IActionResult 的子类，则返回 returnValue，否则调用一个 Convert 方法转换 returnValue：

```
public IActionResult Convert(object value, Type returnType)
{
    if (returnType == null)
    {
        throw new ArgumentNullException(nameof(returnType));
    }

    if (value is IConvertToActionResult converter)
    {
        return converter.Convert();
    }

    return new ObjectResult(value)
    {
        DeclaredType = returnType,
    };
}
```

这个方法会判断 returnValue 是否实现了 IConvertToActionResult 接口，如果是，则调用该接口的 Convert 方法转换成 IActionResult 类型，否则会将 returnValue 封装成 ObjectResult。ObjectResult 也是 ActionResult 的子类。

系统设置了多种 *XXX*ResultExecutor 来处理不同类型的 Action，最终结果无论是什么，都会被转换成 IActionResult 类型，以便在图 20-1 所示的第三部分执行 IActionResult。

20.1 节列出了多种不同类型的 Action，下面通过图 20-2 来看它们的处理结果。

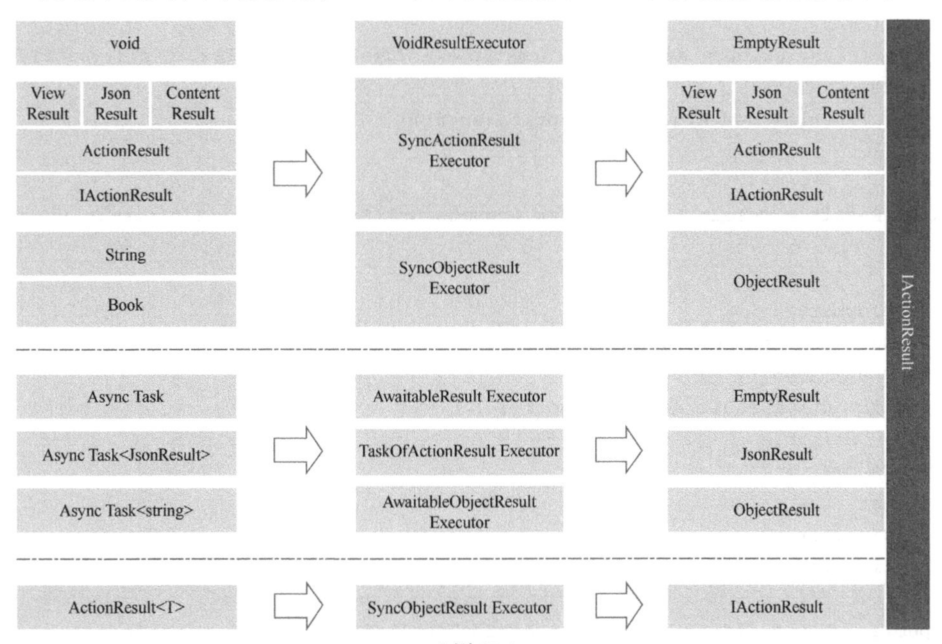

▲图 20-2

void 类型本身没有返回结果，但它会被赋予一个结果 EmptyResult，也是 ActionResult 的子类。

图 20-2 被两条虚线分隔为 3 行，第一行基本介绍过了，第二行是第一行对应的异步方法，这些异步方法的返回结果和对应的同步方法是一样的。由图 20-2 可知，处理它们的 *XXX*ResultExecutor 方法是不一样的。

第三行的 ActionResult<T> 类型是在 ASP.NET Core 2.1 中引入的，它支持 IActionResult 的子类，也支持类似 string 和 Book 等特定类型。

```
public sealed class ActionResult<TValue> : IConvertToActionResult
{
    public ActionResult(TValue value)
    {
        if (typeof(IActionResult).IsAssignableFrom(typeof(TValue)))
        {
            var error = Resources.FormatInvalidTypeTForActionResultOfT(typeof(TValue),
"ActionResult<T>");
            throw new ArgumentException(error);
        }

        Value = value;
    }

    public ActionResult(ActionResult result)
    {
        if (typeof(IActionResult).IsAssignableFrom(typeof(TValue)))
        {
            var error = Resources.FormatInvalidTypeTForActionResultOfT(typeof(TValue),
"ActionResult<T>");
            throw new ArgumentException(error);
        }

        Result = result ?? throw new ArgumentNullException(nameof(result));
    }

    // <summary>
    // Gets the <see cref="ActionResult"/>.
    // </summary>
    public ActionResult Result { get; }

    // <summary>
    // Gets the value.
    // </summary>
    public TValue Value { get; }

    public static implicit operator ActionResult<TValue>(TValue value)
    {
```

```
        return new ActionResult<TValue>(value);
    }

    public static implicit operator ActionResult<TValue>(ActionResult result)
    {
        return new ActionResult<TValue>(result);
    }

    IActionResult IConvertToActionResult.Convert()
    {
        return Result ?? new ObjectResult(Value)
        {
            DeclaredType = typeof(TValue),
        };
    }
}
```

TValue 不支持 IActionResult 及其子类，值若是 IActionResult 子类，会被赋值给 Result 属性，否则会赋值给 Value 属性。它实现了 IConvertToActionResult 接口，当返回结果实现了 IConvertToActionResult 接口时，就会调用 Convert 方法进行转换。Convert 方法先判断值是否是 IActionResult 的子类，如果是，则返回该值；否则将该值转换为 ObjectResult 后返回。

所以图 20-2 中 ActionResult<T> 类型返回的结果被加上引号，意思是结果类型可能是直接返回的 IActionResult 的子类，也可能是 string 和 Book 这样的特定类型被封装后的 ObjectResult 类型。

20.2.3 Result Filter 的执行

结果被统一处理为 IActionResult 后，进入图 20-1 的第三部分，这里主要有两部分内容，分别是 Result Filter 的执行和 IActionResult 的执行。Result Filter 有 OnResultExecuting 和 OnResultExecuted 两个方法，分别在 IActionResult 执行的前后执行。

自定义一个 MyResultFilterAttribute，代码如下：

```
public class MyResultFilterAttribute : Attribute, IResultFilter
{
    public void OnResultExecuted(ResultExecutedContext context)
    {
        Debug.WriteLine("HomeController=======>OnResultExecuted");
    }

    public void OnResultExecuting(ResultExecutingContext context)
    {
        Debug.WriteLine("HomeController=======>OnResultExecuting");
    }
}
```

将它注册到 20.1 节 JSON 的例子中：

```
[MyResultFilter]
public JsonResult GetJson()
{
    return new JsonResult(new Book() { Code ="1001", Name = "ASP" });
}
```

输出结果如下：

```
HomeController=======>OnResultExecuting
……Executing JsonResult……
HomeController=======>OnResultExecuted
```

在 OnResultExecuting 中可以通过设置 context.Cancel = true 取消后面代码的执行：

```
public void OnResultExecuting(ResultExecutingContext context)
{
    //省略用于验证的代码
    context.Cancel = true;
    Debug.WriteLine("HomeController=======>OnResultExecuting");
}
```

输出结果如下：

```
HomeController=======>OnResultExecuting
```

同时返回结果也不再是 JSON 值，返回结果以及 Content-Type 全部为空，IactionResult 和 OnResultExecuted 都不再被执行。

这里除了可以通过 IActionResult 执行之前的验证，还可以对 HttpContext.Response 做一些简单的操作，例如添加一个 Header 值：

```
public void OnResultExecuting(ResultExecutingContext context)
{
    context.HttpContext.Response.Headers.Add("version", "1.2");
    Debug.WriteLine("HomeController=======>OnResultExecuting");
}
```

除了正常返回 JSON 值外，Header 中会出现新添加的 version：

```
Content-Type: application/json; charset=utf-8
version: 1.2
```

OnResultExecuted 方法是在 IActionResult 执行之后执行。因为在这个方法执行时，请求结果已经发送给请求的客户端了，所以这里可以做一些日志类的操作。假如在这个方法中发生了异常：

```
public void OnResultExecuted(ResultExecutedContext context)
{
    throw new Exception("exception");
    Debug.WriteLine("HomeController=======>OnResultExecuted");
}
```

请求结果依然会返回正常的 JSON，但与输出结果中看到的是不一样的：

```
HomeController=======>OnResultExecuting
……
System.Exception: exception
```

发生异常，后面的 Debug 输出没有执行，却将正确的结果返回给了客户端。

20.2.4 IActionResult 的执行

在 ResourceInvoker 的 case State.ResultInside 阶段会调用 IActionResult 的执行方法 InvokeResultAsync。该方法中参数 IActionResult result 会被调用的 ExecuteResultAsync 方法执行。

```
protected virtual async Task InvokeResultAsync(IActionResult result)
{
    var actionContext = _actionContext;

    _diagnosticSource.BeforeActionResult(actionContext, result);
    _logger.BeforeExecutingActionResult(result);

    try
    {
        await result.ExecuteResultAsync(actionContext);
    }
    finally
    {
        _diagnosticSource.AfterActionResult(actionContext, result);
        _logger.AfterExecutingActionResult(result);
    }
}
```

由图 20-2 可知，虽然所有类型的 Action 的结果都被转换成了 IActionResult，但它们在本质上还是有区别的。所以 IActionResult 类型的参数 result 实际上可能是 JsonResult、ViewResult、EmptyResult 等具体类型。下面依然以 20.1 节 JSON 为例，它返回了一个 JsonResult，这里会调用 JsonResult 的 ExecuteResultAsync 方法，JsonResult 的代码如下：

```
public class JsonResult : ActionResult, IStatusCodeActionResult
{
//省略部分代码

    public override Task ExecuteResultAsync(ActionContext context)
```

```
        {
            if (context == null)
            {
                throw new ArgumentNullException(nameof(context));
            }

            var services = context.HttpContext.RequestServices;
            var executor = services.GetRequiredService<JsonResultExecutor>();
            return executor.ExecuteAsync(context, this);
        }
    }
```

在 ExecuteResultAsync 方法中会获取依赖注入中设置的 JsonResultExecutor，由 JsonResultExecutor 来调用 ExecuteAsync 方法执行后面的工作。JsonResultExecutor 的代码如下：

```
public class JsonResultExecutor
{
//省略部分代码

    public virtual async Task ExecuteAsync(ActionContext context, JsonResult result)
    {
        //省略验证代码

        var response = context.HttpContext.Response;

        ResponseContentTypeHelper.ResolveContentTypeAndEncoding(
            result.ContentType,
            response.ContentType,
            DefaultContentType,
            out var resolvedContentType,
            out var resolvedContentTypeEncoding);

        response.ContentType = resolvedContentType;

        if (result.StatusCode != null)
        {
            response.StatusCode = result.StatusCode.Value;
        }

        var serializerSettings = result.SerializerSettings ?? Options.SerializerSettings;

        Logger.JsonResultExecuting(result.Value);
        using (var writer = WriterFactory.CreateWriter(response.Body,
resolvedContentTypeEncoding))
        {
            using (var jsonWriter = new JsonTextWriter(writer))
```

```
                {
                    jsonWriter.ArrayPool = _charPool;
                    jsonWriter.CloseOutput = false;
                    jsonWriter.AutoCompleteOnClose = false;

                    var jsonSerializer = JsonSerializer.Create(serializerSettings);
                    jsonSerializer.Serialize(jsonWriter, result.Value);
                }

                await writer.FlushAsync();
            }
        }
    }
```

JsonResultExecutor 的 ExecuteAsync 方法的作用是将 JsonResult 中的值转换成 JSON 并写入 context.HttpContext.Response. Body。至此，JsonResult 执行完毕。

ViewResult 会由对应的 ViewExecutor 来执行，通过相应的规则生成一个 HTML 页面。

EmptyResult 的 ExecuteResult 方法为空，不会返回任何内容。

```
    public class EmptyResult : ActionResult
    {
        // <inheritdoc />
        public override void ExecuteResult(ActionContext context)
        {
        }
    }
```

以上几种类型的返回结果的格式是固定的，JsonResult 会返回 JSON 格式，ViewResult 会返回 HTML 格式。由 20.1 节的例子可知，string 类型会返回 string 类型的字符串，而实体类型 Book 却会返回 JSON。由图 20-2 可知，这两种类型在执行完毕后都被封装成了 ObjectResult，那么 ObjectResult 在执行时是如何被转换成 string 和 JSON 格式的呢？

20.2.5　ObjectResult 的执行与返回格式的协商

依然以返回 Book 类型的 Action 为例，来看它是怎么被转换为 JSON 类型的。代码如下：

```
    public Book GetModel()
    {
        return new Book() { Code = "1001", Name = "ASP" };
}
```

Action 执行后被封装为 ObjectResult，接下来是 ObjectResult 的执行过程。

ObjectResult 的代码如下：

```
public class ObjectResult : ActionResult, IStatusCodeActionResult
    {
```

```
    //省略部分代码

        public override Task ExecuteResultAsync(ActionContext context)
        {
            var executor = context.HttpContext.RequestServices.GetRequiredService
<IActionResultExecutor<ObjectResult>>();
            return executor.ExecuteAsync(context, this);
        }
    }
```

首先通过依赖注入获取 ObjectResult 对应的执行者，获取 ObjectResultExecutor，然后调用 ObjectResultExecutor 的 ExecuteAsync 方法。代码如下：

```
public class ObjectResultExecutor : IActionResultExecutor<ObjectResult>
{
    //省略部分代码

    public virtual Task ExecuteAsync(ActionContext context, ObjectResult result)
    {
        //省略部分代码

        var formatterContext = new OutputFormatterWriteContext(
            context.HttpContext,
            WriterFactory,
            objectType,
            result.Value);

        var selectedFormatter = FormatterSelector.SelectFormatter(
            formatterContext,
            (IList<IOutputFormatter>)result.Formatters ?? Array.Empty<IOutputFormatter>(),
            result.ContentTypes);
        if (selectedFormatter == null)
        {
            // 未找到适合此格式的 Formatter，记录日志并返回 406 错误
            Logger.NoFormatter(formatterContext);

            context.HttpContext.Response.StatusCode = StatusCodes.Status406Not
Acceptable;
            return Task.CompletedTask;
        }

        result.OnFormatting(context);
        return selectedFormatter.WriteAsync(formatterContext);
    }
}
```

核心代码是 FormatterSelector.SelectFormatter 方法，作用是选择一个合适的 Formatter。

Formatter 是一个用于格式化数据的类。系统默认提供了 4 种 Formatter，如图 20-3 所示。

[0]	{Microsoft.AspNetCore.Mvc.Formatters.HttpNoContentOutputFormatter}
[1]	{Microsoft.AspNetCore.Mvc.Formatters.StringOutputFormatter}
[2]	{Microsoft.AspNetCore.Mvc.Formatters.StreamOutputFormatter}
[3]	{Microsoft.AspNetCore.Mvc.Formatters.JsonOutputFormatter}
Raw View	

▲图 20-3

它们都实现了 IOutputFormatter 接口，继承关系如图 20-4 所示。

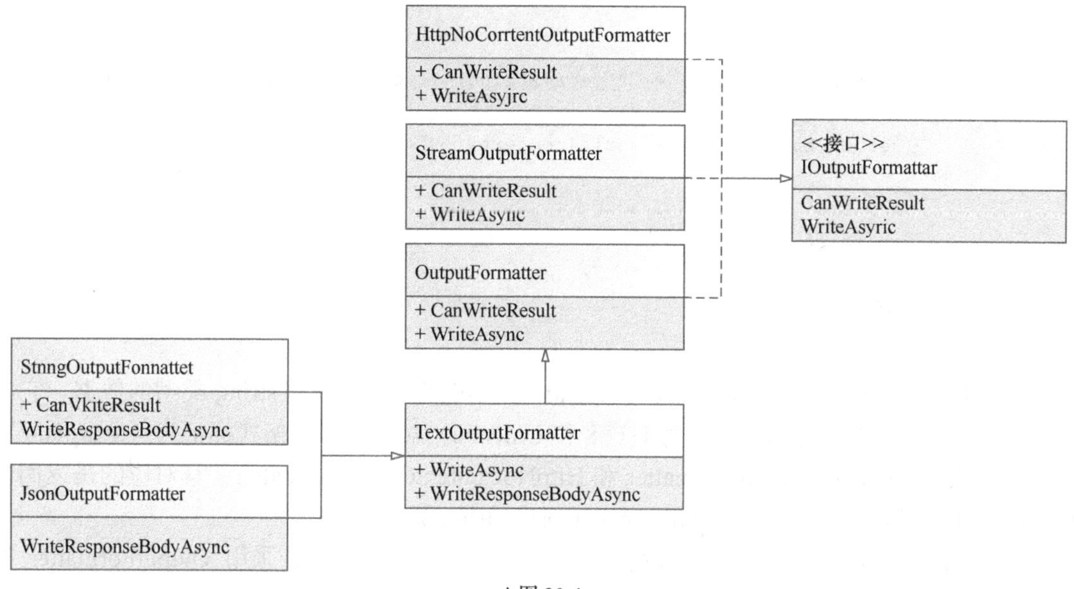

▲图 20-4

IOutputFormatter 的代码如下：

```
public interface IOutputFormatter
{
        bool CanWriteResult(OutputFormatterCanWriteContext context);
        Task WriteAsync(OutputFormatterWriteContext context);
}
```

就像在众多 *XXX*ResultExecutor 中筛选一个合适的 Action 的执行者一样，首先将它们按照一定的顺序排列，然后开始遍历，逐一执行它们的 Can*XXX* 方法，若其中一个的执行结果为 true，则会被选出来。StringOutputFormatter 的代码如下：

```
public class StringOutputFormatter : TextOutputFormatter
{
    public StringOutputFormatter()
    {
```

```
        SupportedEncodings.Add(Encoding.UTF8);
        SupportedEncodings.Add(Encoding.Unicode);
        SupportedMediaTypes.Add("text/plain");
    }

    public override bool CanWriteResult(OutputFormatterCanWriteContext context)
    {
        if (context == null)
        {
            throw new ArgumentNullException(nameof(context));
        }

        if (context.ObjectType == typeof(string) || context.Object is string)
        {
            return base.CanWriteResult(context);
        }

        return false;
    }
    //省略部分代码
}
```

从 StringOutputFormatter 的 CanWriteResult 方法可知，它能处理 string 类型的数据。它的构造方法中标识可以处理的字符集为 UTF8 和 Unicode。对应的数据格式标记为“text/plain”。同样查看 HttpNoContentOutputFormatter 和 HttpNoContentOutputFormatter，其对应的是返回值为 void 或 Task 类型，StreamOutputFormatter 对应 string 类型。

SystemTextJsonOutputFormatter 没有重写 CanWriteResult 方法，采用 OutputFormatter 的 CanWriteResult 方法，代码如下：

```
public abstract class OutputFormatter : IOutputFormatter, IApiResponseTypeMetadataProvider
{
//省略部分代码

    protected virtual bool CanWriteType(Type type)
    {
        return true;
    }

    // <inheritdoc />
    public virtual bool CanWriteResult(OutputFormatterCanWriteContext context)
    {
        if (SupportedMediaTypes.Count == 0)
        {
            var message = Resources.FormatFormatter_NoMediaTypes(
                GetType().FullName,
```

```
                nameof(SupportedMediaTypes));

            throw new InvalidOperationException(message);
        }

        if (!CanWriteType(context.ObjectType))
        {
            return false;
        }

        if (!context.ContentType.HasValue)
        {
            context.ContentType = new StringSegment(SupportedMediaTypes[0]);
            return true;
        }
        else
        {
            var parsedContentType = new MediaType(context.ContentType);
            for (var i = 0; i < SupportedMediaTypes.Count; i++)
            {
                var supportedMediaType = new MediaType(SupportedMediaTypes[i]);
                if (supportedMediaType.HasWildcard)
                {
                    if (context.ContentTypeIsServerDefined
                        && parsedContentType.IsSubsetOf(supportedMediaType))
                    {
                        return true;
                    }
                }
                else
                {
                    if (supportedMediaType.IsSubsetOf(parsedContentType))
                    {
                        context.ContentType = new StringSegment(SupportedMediaTypes[i]);
                        return true;
                    }
                }
            }
        }

        return false;
    }
}
```

此段代码主要是利用 SupportedMediaTypes 和 context.ContentType 做一系列的判断，分别来自客户端和服务端。

- SupportedMediaTypes：由客户端在请求时给出，标识客户端期望服务端按照什么格式返回请求结果。
- context.ContentType：来自 ObjectResult.ContentTypes，由服务端在 Action 执行后给出。

二者的值都类似 application/json text/plain 的格式，也有可能为空，即客户端或服务端未对请求做数据格式的设定。通过上面的代码可知，如果这两个值均未做设置或者只有一方设置为 JSON 时，那么 CanWriteResult 方法的返回值都是 true。所以，除了前 3 种 Formatter 对应的特定类型的 ObjectResult，其余的都会交由 JsonOutputFormatter 处理。这也就是为什么同样是 ObjectResult，但 string 类型的 Action 返回结果是 string 类型，而 Book 类型的 Action 返回的结果是 JSON 类型。JsonOutputFormatter 相当于当其他的 Formatter 无法处理时用来"保底"的。

那么 SupportedMediaTypes 和 context.ContentType 是在什么时候被设置的呢？在请求的模型参数绑定时，可以通过在请求 Request 的 Header 中添加 content-type: application/json 标识来说明请求中包含的数据格式是 JSON 类型的。同样，在请求时也可以添加 accept:*XXX* 标识，表明期望服务端对本次请求返回的数据格式。例如，期望是 JSON 格式用 accept:application/json、文本格式用 accept: text/plain 等，这个值就是 SupportedMediaTypes。

在服务端，可以对返回的数据格式做设置，代码如下：

```
[Produces("application/json")]
public Book GetModel()
{
    return new Book() { Code = "1001", Name = "ASP" };
}
```

ProducesAttribute 设置的值会被赋值给 ObjectResult.ContentTypes，最终传递给 context.ContentType。ProducesAttribute 实际是一个 IResultFilter，代码如下：

```
public class ProducesAttribute : Attribute, IResultFilter, IOrderedFilter, IApiRespon
seMetadataProvider
    {
        //省略部分代码

        public virtual void OnResultExecuting(ResultExecutingContext context)
        {
            //省略部分代码

            SetContentTypes(objectResult.ContentTypes);
        }

        public void SetContentTypes(MediaTypeCollection contentTypes)
        {
            contentTypes.Clear();
            foreach (var contentType in ContentTypes)
            {
```

```
                contentTypes.Add(contentType);
            }
        }

        private MediaTypeCollection GetContentTypes(string firstArg, string[] args)
        {
            var completeArgs = new List<string>();
            completeArgs.Add(firstArg);
            completeArgs.AddRange(args);
            var contentTypes = new MediaTypeCollection();
            foreach (var arg in completeArgs)
            {
                var contentType = new MediaType(arg);
                if (contentType.HasWildcard)
                {
                    throw new InvalidOperationException(
                        Resources.FormatMatchAllContentTypeIsNotAllowed(arg));
                }

                contentTypes.Add(arg);
            }

            return contentTypes;
        }
    }
```

在执行 OnResultExecuting 时，会将设置的 application/json 赋值给 ObjectResult.ContentTypes，所以请求最终返回的结果的数据格式是由二者“协商”决定的。下面回到 Formatter 的筛选方法 FormatterSelector.SelectFormatter()，这个方法写在 DefaultOutputFormatterSelector.cs 中。精简后的代码如下：

```
public class DefaultOutputFormatterSelector : OutputFormatterSelector
{
    //省略部分代码

    public override IOutputFormatter SelectFormatter(OutputFormatterCanWriteContext
context, IList<IOutputFormatter> formatters, MediaTypeCollection contentTypes)
    {
        //省略部分代码

        var request = context.HttpContext.Request;
        var acceptableMediaTypes = GetAcceptableMediaTypes(request);
        var selectFormatterWithoutRegardingAcceptHeader = false;

        IOutputFormatter selectedFormatter = null;
        if (acceptableMediaTypes.Count == 0)
```

```
    {
        //客户端未设置 Accept 标识的情况
        selectFormatterWithoutRegardingAcceptHeader = true;
    }
    else
    {
        if (contentTypes.Count == 0)
        {
            //服务端未指定数据格式的情况
            selectedFormatter = SelectFormatterUsingSortedAcceptHeaders(
                context,
                formatters,
                acceptableMediaTypes);
        }
        else
        {
            //客户端和服务端均指定了数据格式的情况
            selectedFormatter = SelectFormatterUsingSortedAcceptHeadersAndContent
Types(
                context,
                formatters,
                acceptableMediaTypes,
                contentTypes);
        }

        if (selectedFormatter == null)
        {
            //如果未找到合适的，由系统参数 ReturnHttpNotAcceptable 决定直接返回错误
            //还是忽略客户端的 Accept 设置再筛选一次
            if (!_returnHttpNotAcceptable)
            {
                selectFormatterWithoutRegardingAcceptHeader = true;
            }
        }
    }

    if (selectFormatterWithoutRegardingAcceptHeader)
    {
        //Accept 未设置或者被忽略的情况
        if (contentTypes.Count == 0)
        {
            //服务端也未指定数据格式的情况
            selectedFormatter = SelectFormatterNotUsingContentType(
                context,
                formatters);
        }
```

```
            else
            {
                //服务端指定数据格式的情况
                selectedFormatter = SelectFormatterUsingAnyAcceptableContentType(
                    context,
                    formatters,
                    contentTypes);
            }
        }

        if (selectedFormatter == null)
        {
            // 未找到适合此格式的 formatter
            _logger.NoFormatter(context);
            return null;
        }

        _logger.FormatterSelected(selectedFormatter, context);
        return selectedFormatter;
    }

    // 省略 4 种情况对应的 4 个方法
    // SelectFormatterNotUsingContentType
    // SelectFormatterUsingSortedAcceptHeaders
    // SelectFormatterUsingAnyAcceptableContentType
    // SelectFormatterUsingSortedAcceptHeadersAndContentTypes

}
```

DefaultOutputFormatterSelector 根据客户端和服务端关于返回数据格式而设置的 4 种不同情况分别做了处理，优化了查找顺序，此处就不详细讲解了。

总结规则如下。

（1）只有在 Action 返回类型为 ObjectResult 时才会进行“协商”。如果返回类型为 JsonResult、ContentResult、ViewResult 等特定 ActionResult，无论请求是否设置了 Accept 标识，都会被忽略，固定返回 JSON、string、HTML 类型的结果。

（2）当系统检测到请求来自客户端时，会忽略其 Header 中 Accept 的设置，由服务端设置的格式决定（未做特殊配置时，系统默认为 JSON），这是为了在不同客户端使用 API 时提供更一致的体验。系统提供了参数 RespectBrowserAcceptHeader，即接受客户端在请求的 Header 中关于 Accept 的设置，默认值为 false。将其设置为 true 时，客户端请求中的 Accept 标识才会生效。注意，这只是使该 Accept 标识生效，依然不能由其决定返回格式，会进入“协商”阶段。

（3）若二者均未设置，则采用默认的 JSON 格式。

（4）若二者其中有一个被设置，则采用该设置值。

（5）若二者均已设置且不一致，即二者值不相同且没有包含关系（有通配符的情况），会判断系统参数 ReturnHttpNotAcceptable（返回不可接受，默认值为 false）。若 ReturnHttpNotAcceptable 值为 false，则忽略客户端的 Accept 设置，按照无 Accept 设置的情况再筛选一次 Formatter；如果该值为 true，则直接返回状态 406。

涉及两个系统参数 RespectBrowserAcceptHeader 和 ReturnHttpNotAcceptable 的设置方法是在 Startup.cs 中通过如下代码设置：

```
services.AddMvc(
    options =>
    {
        options.RespectBrowserAcceptHeader = true;
options.ReturnHttpNotAcceptable = true;
    }
)
```

最终，通过上述方法找到合适的 Formatter，接着通过该 Formatter 的 WriteAsync 方法将请求结果格式化后写入 HttpContext.Response。SystemTextJsonOutputFormatter 重写了 OutputFormatter 的 WriteResponseBodyAsync 方法（WriteAsync 方法会调用 WriteResponseBodyAsync 方法），代码如下：

```
public sealed override async Task WriteResponseBodyAsync(OutputFormatterWriteContext
context, Encoding selectedEncoding)
{
    var httpContext = context.HttpContext;

    var writeStream = GetWriteStream(httpContext, selectedEncoding);
    try
    {
        var objectType = context.Object?.GetType() ?? context.ObjectType;
        await JsonSerializer.SerializeAsync(writeStream, context.Object, objectType,
SerializerOptions);

        if (writeStream is TranscodingWriteStream transcodingStream)
        {
            await transcodingStream.FinalWriteAsync(CancellationToken.None);
        }
        await writeStream.FlushAsync();
    }
    finally
    {
        if (writeStream is TranscodingWriteStream transcodingStream)
        {
            await transcodingStream.DisposeAsync();
        }
```

```
    }
}
```

这个方法的功能是将结果数据转换为 JSON 并写入 HttpContext.Response. Body。至此，请求结果就以 JSON 格式返回给客户端了。

在实际项目中，如果上述的几种格式均不能满足需求，比如某种数据经常需要通过特殊的格式传输，想自定义一种格式，那么有两种方式来实现，即自定义 IActionResult 或 IOutputFormatter。

20.3 自定义 IActionResult

以 20.1 节的第 3 个例子为例，该例通过 return new JsonResult(new Book() { Code = "1001", Name = "ASP" })返回了一个 JsonResult。

返回的 JSON 值为：

```
{"code":"1001","name":"ASP"}
```

假如希望用特殊的格式返回 Book 类型，例如以下格式：

```
Book Code:[1001]|Book Name:<ASP>
```

可以通过自定义一个类似 JsonResult 的类来实现。代码如下：

```
public class BookResult : ActionResult
    {
        public BookResult(Book content)
        {
            Content = content;
        }
        public Book Content { get; set; }
        public string ContentType { get; set; }
        public int? StatusCode { get; set; }

        public override async Task ExecuteResultAsync(ActionContext context)
        {
            if (context == null)
            {
                throw new ArgumentNullException(nameof(context));
            }

            var executor = context.HttpContext.RequestServices.GetRequiredService
<IActionResultExecutor<BookResult>>();
            await executor.ExecuteAsync(context, this);
        }


    }
```

为了方便继承 ActionResult，定义了一个名为 BookResult 的类。为了处理 Book 类型，在构造方法中添加了 Book 类型的参数，并将该参数赋值给 Content 属性。重写 ExecuteResultAsync 方法，对应 JsonResultExecutor，还需要自定义一个 BookResultExecutor。代码如下：

```
public class BookResultExecutor : IActionResultExecutor<BookResult>
{
    private const string DefaultContentType = "text/plain; charset=utf-8";
    private readonly IHttpResponseStreamWriterFactory _httpResponseStreamWriterFactory;

    public BookResultExecutor(IHttpResponseStreamWriterFactory httpResponseStreamWriterFactory)
    {
    _httpResponseStreamWriterFactory = httpResponseStreamWriterFactory;
    }
    private static string FormatToString(Book book)
    {
        return string.Format("Book Code:[{0}]|Book Name:<{1}>", book.Code, book.Name);
    }

    // <inheritdoc />
    public virtual async Task ExecuteAsync(ActionContext context, BookResult result)
    {
        if (context == null)
        {
            throw new ArgumentNullException(nameof(context));
        }

        if (result == null)
        {
            throw new ArgumentNullException(nameof(result));
        }

        var response = context.HttpContext.Response;

        string resolvedContentType;
        Encoding resolvedContentTypeEncoding;
        ResponseContentTypeHelper.ResolveContentTypeAndEncoding(
            result.ContentType,
            response.ContentType,
            DefaultContentType,
            out resolvedContentType,
            out resolvedContentTypeEncoding);

        response.ContentType = resolvedContentType;

        if (result.StatusCode != null)
        {
```

```
            response.StatusCode = result.StatusCode.Value;
        }

        string content = FormatToString(result.Content);

        if (result.Content != null)
        {
            response.ContentLength = resolvedContentTypeEncoding.GetByteCount(content);

            using (var textWriter = _httpResponseStreamWriterFactory.CreateWriter
(response.Body, resolvedContentTypeEncoding))
            {
                await textWriter.WriteAsync(content);

                await textWriter.FlushAsync();
            }
        }
    }
}
```

这里定义了默认的 ContentType 类型，采用文本格式，即"text/plain; charset=utf-8"，这会在请求结果的 Header 中出现。为了说明特殊格式，也可以自定义一个特殊类型，例如"text/book; charset=utf-8"，这需要在项目开头约定好。定义一个 FormatToString 方法，用于将 Book 类型的数据格式化，最终将格式化的数据写入 Response.Body。

定义 BookResultExecutor 后，需要在依赖注入（Startup 中的 ConfigureServices 方法）中注册：

```
public void ConfigureServices(IServiceCollection services)
{
    //省略部分代码

    services.TryAddSingleton<IActionResultExecutor<BookResult>, BookResultExecutor>();
}
```

至此，自定义的 BookResult 就可以被使用了，例如下面代码所示的 Action：

```
public BookResult GetBookResult()
{
    return new BookResult(new Book() { Code = "1001", Name = "ASP" });
}
```

用 Fiddler 访问 Action 并测试，返回结果如下：

```
Book Code:[1001]|Book Name:<ASP>
```

Header 中的标识值：

```
Content-Length: 32
Content-Type: text/book; charset=utf-8
```

这是自定义了 ContentType 的结果。

20.4 自定义格式化类

对于 20.3 节的例子，也可以对照 JsonOutputFormatter 自定义一个格式化类来实现。新定义一个名为 BookOutputFormatter 的类，也如同 JsonOutputFormatter 一样继承 TextOutputFormatter。代码如下：

```
public class BookOutputFormatter : TextOutputFormatter
    {
        public BookOutputFormatter()
        {
            SupportedEncodings.Add(Encoding.UTF8);
            SupportedEncodings.Add(Encoding.Unicode);
            SupportedMediaTypes.Add("text/book");
        }

        public override bool CanWriteResult(OutputFormatterCanWriteContext context)
        {
            if (context == null)
            {
                throw new ArgumentNullException(nameof(context));
            }
            if (context.ObjectType == typeof(Book) || context.Object is Book)
            {
                return base.CanWriteResult(context);
            }

            return false;
        }

        private static string FormatToString(Book book)
        {
            return string.Format("Book Code:[{0}]|Book Name:<{1}>",book.Code,book.Name);
        }


        public override async Task WriteResponseBodyAsync(OutputFormatterWrite
Context context, Encoding selectedEncoding)
        {


            if (context == null)
            {
```

```
                throw new ArgumentNullException(nameof(context));
            }

            if (selectedEncoding == null)
            {
                throw new ArgumentNullException(nameof(selectedEncoding));
            }

            var valueAsString = FormatToString(context.Object as Book);
            if (string.IsNullOrEmpty(valueAsString))
            {
                await Task.CompletedTask;
            }

            var response = context.HttpContext.Response;
            await response.WriteAsync(valueAsString, selectedEncoding);
        }
    }
```

首先在构造方法中定义它所支持的字符集和ContentType类型。重写CanWriteResult方法，用于确定它能处理对应的请求返回结果。可以在此方法中做多种判断，最终返回bool类型的结果。本例比较简单，仅判断返回的结果是否为Book类型，同样定义了FormatToString方法用于请求结果的格式化。最后重写WriteResponseBodyAsync方法，将格式化后的结果写入Response.Body。

BookOutputFormatter定义之后也需要注册到系统中，例如以下代码：

```
services.AddMvc(
   options =>
   {
       options.OutputFormatters.Insert(0,new BookOutputFormatter());
   }
)
```

这里采用了Insert方法，也就是将其插入了OutputFormatters集合的第一个Formatter。所以在筛选OutputFormatters时，它也是第一个。此时的OutputFormatters集合如图20-5所示。

[0]	{IntOutPutFormatterDemo.BookOutputFormatter}
[1]	{Microsoft.AspNetCore.Mvc.Formatters.HttpNoContentOutputFormatter}
[2]	{Microsoft.AspNetCore.Mvc.Formatters.StringOutputFormatter}
[3]	{Microsoft.AspNetCore.Mvc.Formatters.StreamOutputFormatter}
[4]	{Microsoft.AspNetCore.Mvc.Formatters.JsonOutputFormatter}
Raw View	

▲图20-5

通过Fiddler进行测试，以20.1节返回Book类型的第4个例子为例：

```
   public Book GetModel()
   {
       return new Book() { Code = "1001", Name = "ASP" };
}
```

当设定 accept: text/book 或者未设定 Accept 时，采用自定义的 BookOutputFormatter，返回结果为：

```
Book Code:[1001]|Book Name:<ASP>
```

ContentType 值为：

```
text/book; charset=utf-8
```

当设定"accept: application/json"时，返回 JSON，值为：

```
{"code":"1001","name":"ASP"}
```

ContentType 值为：

```
application/json; charset=utf-8
```

这是由于 BookOutputFormatter 类型排在了 JsonOutputFormatter 的前面，所以对于 Book 类型会首先采用 BookOutputFormatter；当客户端通过 Accept 方式要求返回结果为 JSON 时，才采用 JSON 类型。测试服务端的要求，为 Action 添加 Produces 设置，代码如下：

```
[Produces("application/json")]
public Book GetModel()
{
    return new Book() { Code = "1001", Name = "ASP" };
}
```

此时无论是设定 accept: text/book，还是未设定 Accept 的情况，都会按照 JSON 类型返回结果。这也验证了 20.2 节关于服务端和客户端"协商"的规则。

20.5 添加 XML 类型支持

20.3 节和 20.4 节通过自定义的方式实现了特殊格式的处理，在项目中常见的格式还有 XML，在 ASP.NET Core 中没有做默认支持。如果需要 XML 格式的支持，可以通过 NuGet 添加相应的包。

在 NuGet 中搜索并安装 Microsoft.AspNetCore.Mvc.Formatters.Xml，如图 20-6 所示。

▲图 20-6

不需要使用像 BookOutputFormatter 的注册方式。系统提供了注册方法：

```
services.AddMvc().AddXmlSerializerFormatters();
```

或者：

```
services.AddMvc().AddXmlDataContractSerializerFormatters();
```

分别对应了两种格式化程序：System.Xml.Serialization.XmlSerializer 和 System.Runtime.Serialization.DataContractSerializer。

二者的区别就不在这里描述了。注册之后，可以通过在请求的 Header 中设置 accept: application/xml 来获取 XML 的结果。访问 20.4 节的返回结果的数据类型为 Book 的例子，返回结果如下：

```
<Book xmlns:xsi="http://www.w3.org/2001/XMLSchema-instance" xmlns:xsd="http://www.w3.org/2001/XMLSchema">
    <Code>1001</Code>
    <Name>ASP</Name>
</Book>
```

ContentType 值为：

```
application/xml; charset=utf-8
```

第 21 章　一个 API 与小程序的项目

本章通过一个简单的例子来回顾前面章节的知识点。

21.1 前期准备

本项目是一个用于记录宝宝成长的微信小程序，相关系统及环境如下。

- ❑ **客户端**：微信小程序。
- ❑ **服务端**：ASP.NET Core 3 的 API 项目。
- ❑ **服务端操作系统**：CentOS 7。
- ❑ **数据库**：MongoDB。

主要功能是记录宝宝不同时期的身高和体重，可以通过列表和折线图两种方式查看这些数据，主要功能页面如图 21-1 所示。

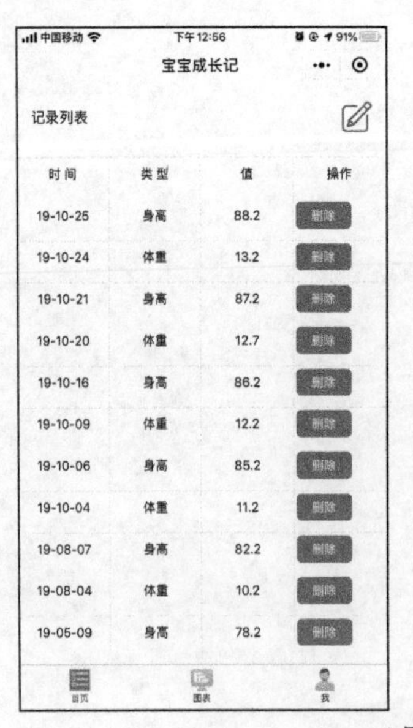

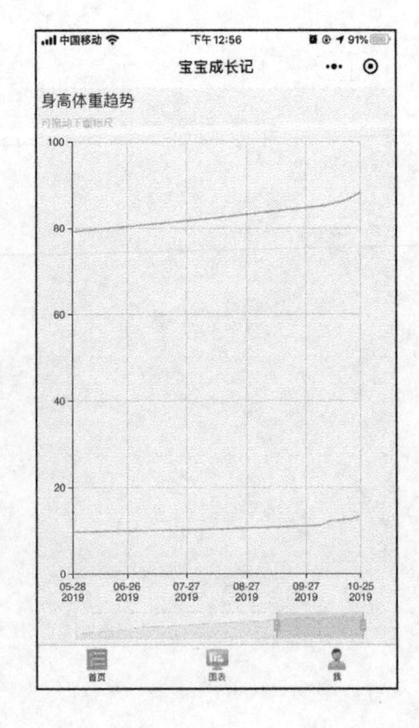

▲图 21-1

21.1.1 服务器环境搭建

服务器环境采用第 6 章配置的环境，按本项目需求再安装 MongoDB 数据库即可。

21.1.2 安装 MongoDB 数据库

（1）创建一个名为/etc/yum.repos.d/mongodb-org-4.2.repo 的文件，代码如下：

```
Vi /etc/yum.repos.d/mongodb-org-4.2.repo
```

文件内容如下：

```
[mongodb-org-4.2]
name=MongoDB Repository
baseurl=https://repo.mongodb.org/yum/redhat/$releasever/mongodb-org/4.2/x86_64/
gpgcheck=1
enabled=1
gpgkey=https://www.mongodb.org/static/pgp/server-4.2.asc
```

（2）执行下面命令开始安装，等待安装完成：

```
sudo yum install -y mongodb-org
```

（3）启动 MongoDB：

```
systemctl start mongod.service
```

（4）验证启动结果：

```
systemctl status mongod.service
```

（5）设为开机启动：

```
systemctl enablemongod.service
```

MongoDB 的默认端口是 27017，不允许外网访问，可以修改/etc/mongod.conf 文件改变此设置。默认情况下该文件的配置如图 21-2 所示。

```
# network interfaces
net:
  port: 27017
  bindIp: 127.0.0.1  # Enter 0.0.0.0,:: to bind to all IPv4 and IPv6 addresses or,
alternatively, use the net.bindIpAll setting.
```

▲图 21-2

如果在开发阶段临时开放外网访问，将 bindIp 修改为 0.0.0.0 即可。端口也可以不采用默认的，改为其他未被占用的端口，增加安全性。不过要记得开放服务器相应的端口。

客户端工具可以选择安装官方的 MongoDB Compass，支持 Windows、Linux、macOS 版本。

21.1.3　微信小程序注册

在微信公众平台进行注册，选择账号类型为“小程序”，如图 21-3 所示。

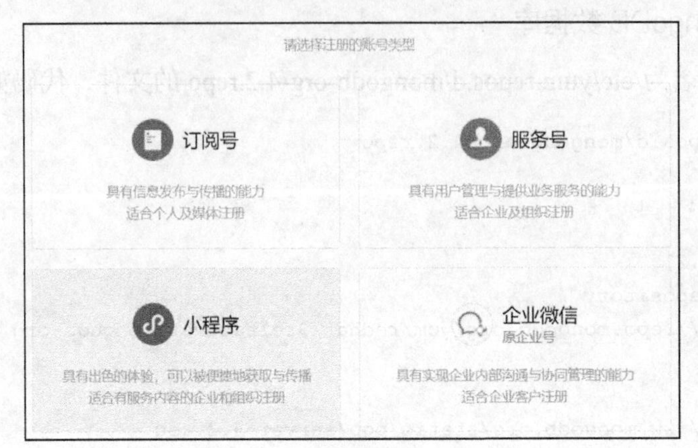

▲图 21-3

接下来按照要求填写相关信息即可，如图 21-4 所示。

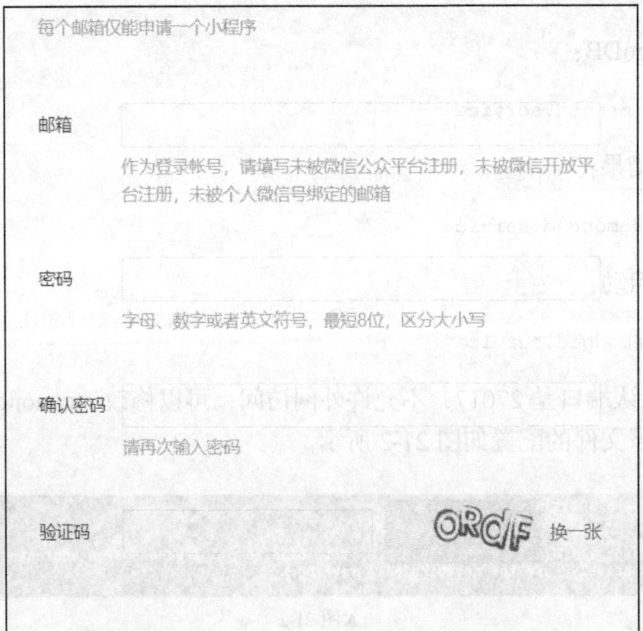

每个邮箱仅能申请一个小程序

邮箱

作为登录帐号，请填写未被微信公众平台注册，未被微信开放平台注册，未被个人微信号绑定的邮箱

密码

字母、数字或者英文符号，最短8位，区分大小写

确认密码

请再次输入密码

验证码　ORCF　换一张

你已阅读并同意《微信公众平台服务协议》及《微信小程序平台服务条款》

注册

▲图 21-4

注册成功后登录，在首页会看到如图 21-5 所示的提示页面。

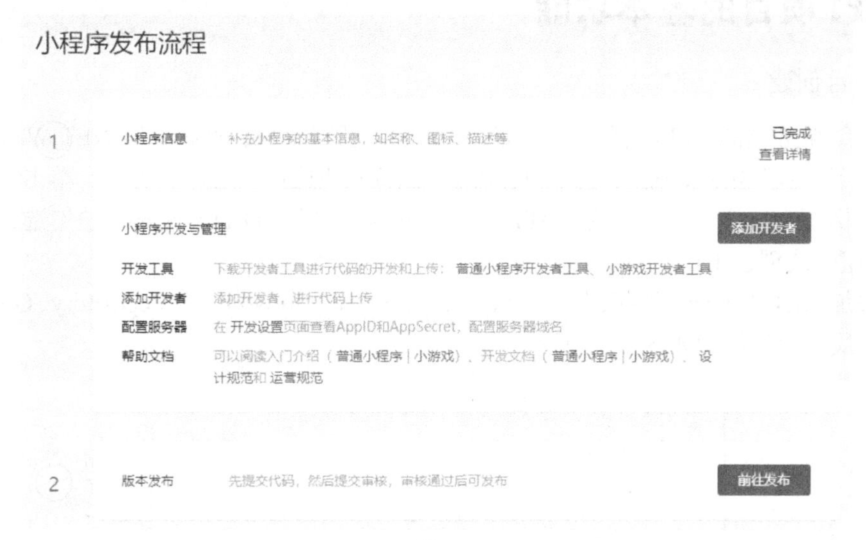

▲图 21-5

其他设置比较简单，主要来看配置服务器项，单击“开发设置”进入配置页面，如图 21-6 所示。

▲图 21-6

对于服务器域名的配置，本项目都采用同样的 api.*XXX*.cn 作为配置值。记录小程序的 AppID 和 AppSecret，在项目中会用到。

21.2 API 项目的基本功能

21.2.1 项目创建

下面介绍服务端的 API 项目。使用 Visual Studio 新建项目，选择 ASP.NET Web 应用程序后，单击“下一步”按钮，输入项目名称 GrowthDiary 并选择项目存放位置，单击“创建”按钮，在新对话框内选择项目类型为“API”，取消对选项“为 HTTPS 配置”的勾选，最后单击“创建”按钮完成创建工作。

依次添加.NET Standard 类库，如 GrowthDiary.Service、GrowthDiary.Repository、GrowthDiary.Model 等，最终项目结构如图 21-7 所示。

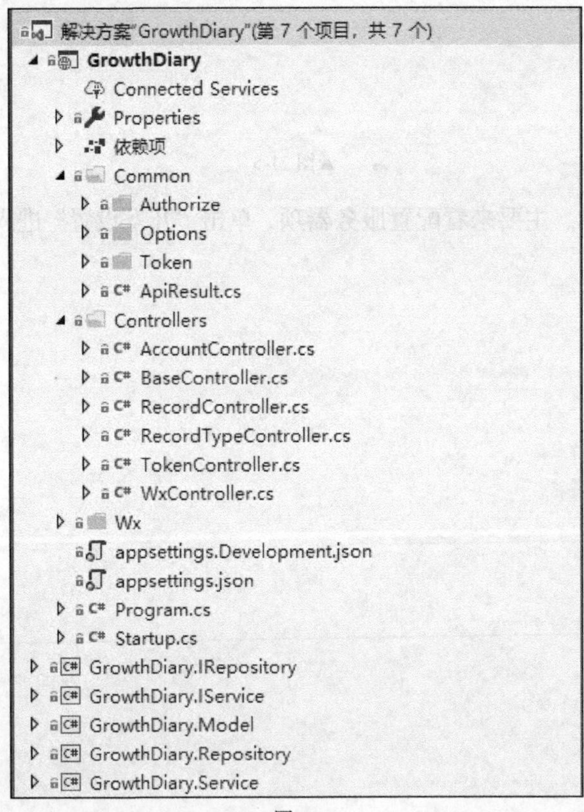

▲图 21-7

21.2.2 操作 MongoDB 数据库

对 MongoDB 的操作的内容放在 GrowthDiary.Repository 层，MongoDB 官方提供了相应的 NuGet 包——MongoDB.Driver。在 GrowthDiary.Repository 层搜索并安装即可，如图 21-8 所示。

▲图 21-8

新建一个 MongoHelper，用于封装对 MongoDB 的操作，代码如下：

```
public class MongoHelper
{
    private readonly IMongoDatabase database;
    public MongoHelper(IConfiguration configuration) : this(configuration["DB:
ConnectionString"], configuration.GetSection("DB:Name").Value)
    {


    }
    public MongoHelper(string ConnectionString, string DBName)
    {
        MongoClient mongoClient = new MongoClient(ConnectionString);
        database = mongoClient.GetDatabase(DBName);
    }

    public T Find<T>(FilterDefinition<T> filter = null, string collectionName = null)
    {
        return FindList<T>(filter, collectionName).FirstOrDefault();
    }
    public List<T> FindList<T>(FilterDefinition<T> filter = null, string collectionName =
 null)
    {
        collectionName ??= typeof(T).Name;
        filter ??= new BsonDocument();
        var collection = database.GetCollection<T>(collectionName);
        return collection.Find(filter).ToList();
    }

    public async Task<T> FindAsync<T>(FilterDefinition<T> filter = null, string
collectionName = null)
    {
        var result = await FindListAsync<T>(filter, collectionName);
        return result.FirstOrDefault();
    }
```

```
    public async Task<List<T>> FindListAsync<T>(FilterDefinition<T> filter = null,
string collectionName = null)
    {
        collectionName ??= typeof(T).Name;
        filter ??= new BsonDocument();
        var collection = database.GetCollection<T>(collectionName);
        var result = await collection.FindAsync(filter);
        return result.ToList();
    }

    public void InsertOne<T>(T model, string collectionName = null)
    {
        collectionName ??= typeof(T).Name;
        var collection = database.GetCollection<T>(collectionName);
        collection.InsertOne(model);
    }

    public async Task InsertOneAsync<T>(T model, string collectionName = null)
    {
        collectionName ??= typeof(T).Name;
        var collection = database.GetCollection<T>(collectionName);
        await collection.InsertOneAsync(model);
    }

    public int UpdateOne<T>(T model, string collectionName = null, params string[]
fields) where T : BaseModel
    {
        var list = new List<UpdateDefinition<T>>();
        bool updateAll = false;
        if (null == fields || fields.Length == 0) updateAll = true;

        foreach (var item in model.GetType().GetProperties())
        {
            if (item.Name.ToLower().Equals("_id")) continue;
            if ((updateAll || fields.Contains(item.Name)))
            {
                list.Add(Builders<T>.Update.Set(item.Name, item.GetValue(model)));
            }
        }

        var updateDefinition = Builders<T>.Update.Combine(list);
        return UpdateOne(model._id, updateDefinition, collectionName);
    }
```

```
    public int UpdateOne<T>(string id, UpdateDefinition<T> updateDefinition, string
collectionName = null) where T : BaseModel
    {
        collectionName ??= typeof(T).Name;
        var collection = database.GetCollection<T>(collectionName);
        var result = collection.UpdateOne(m => m._id == id, updateDefinition);
        return (int)result.ModifiedCount;
    }

    public async Task<int> UpdateOneAsync<T>(T model, string collectionName = null,
params string[] fields) where T : BaseModel
    {
        collectionName ??= typeof(T).Name;
        var collection = database.GetCollection<T>(collectionName);
        var list = new List<UpdateDefinition<T>>();
        bool updateAll = false;
        if (null == fields || fields.Length == 0) updateAll = true;

        foreach (var item in model.GetType().GetProperties())
        {
            if (item.Name.ToLower().Equals("_id")) continue;
            if (updateAll || fields.Contains(item.Name))
            {
                list.Add(Builders<T>.Update.Set(item.Name, item.GetValue(model)));
            }
        }

        var updatefilter = Builders<T>.Update.Combine(list);
        var result = await collection.UpdateOneAsync(m => m._id == model._id,
updatefilter);
        return (int)result.ModifiedCount;
    }

    public int ReplaceOne<T>(T model, string collectionName = null) where T :
BaseModel
    {
        collectionName ??= typeof(T).Name;
        var collection = database.GetCollection<T>(collectionName);
        var result = collection.ReplaceOne(m => m._id == model._id, model);

        return (int)result.ModifiedCount;
    }

    public async Task<int> ReplaceOneAsync<T>(T model, string collectionName = null)
where T : BaseModel
    {
```

```
            collectionName ??= typeof(T).Name;
            var collection = database.GetCollection<T>(collectionName);
            var result = await collection.ReplaceOneAsync(m => m._id == model._id, model);

            return (int)result.ModifiedCount;
        }
    }
```

对 MongoDB.Driver 提供的方法进行封装，以便操作。注意，此处仅对本项目中用到的内容进行了封装，复杂项目中需自行添加其他方法，下文涉及的其他功能和相关的类也是如此。

21.2.3 Model 定义

首先定义两个基类 Model，分别作为所有 Model 和 SearchModel 的基类。

```
public class BaseModel
{
    [BsonRepresentation(BsonType.ObjectId)]
    public string _id { get; set; }
}

public class BaseSearchModel
{
    public bool IsPagination { get; set; } = true;  //是否分页
    public int PageIndex { get; set; } = 1;          //页码
    public int PageSize { get; set; } = 10;          //每页 10 行
    public int RecordCount { get; set; } = 0;        //总行数
}
```

为了对应 MongoDB 的_id 字段，在 Model 层新增一个 Model 的基类 BaseModel，并添加_id 字段。

定义身高和体重记录相关的 Model。

```
    public enum RecordType
    {
        Height = 1,  //身高
        Weight = 2   //体重
    }

    // <summary>
    // 记录类型定义
    // </summary>
    public class RecordTypeDefinition
    {
        // <summary>
        // 记录类型
```

```
        // </summary>
        public RecordType RecordType { get; set; }
        // <summary>
        // 单位
        // </summary>
        public string Unit { get; set; }
        // <summary>
        // 最大值
        // </summary>
        public float MaxValue{ get; set; }
        // <summary>
        // 最小值
        // </summary>
        public float MinValue { get; set; }
    }

    // <summary>
    // 身高、体重记录
    // </summary>
    public class Record : BaseModel
    {
        // <summary>
        // 操作用户 Code
        // </summary>
        public string UserCode { get; set; }
        // <summary>
        // 操作用户 Name
        // </summary>
        public string UserName { get; set; }
        // <summary>
        // 记录类型
        // </summary>
        public int RecordType { get; set; }
        // <summary>
        // 记录值
        // </summary>
        public float Value { get; set; }
        // <summary>
        // 记录状态 默认为 1 正常 2 逻辑删除
        // </summary>
        public int State { get; set; }
        // <summary>
        // 记录时间
        // </summary>
        public DateTime CreateTime { get; set; }
}
```

定义对应的RecordTypeDefinitionViewModel和RecordViewModel。本项目只涉及身高和体重，可将其作为枚举定义，以方便添加其他类型。

21.2.4 Service接口及实现

定义接口基类IBaseService：

```
public interface IBaseService<T> where T: BaseViewModel
{
    Task InsertOneAsync(T viewModel);
    Task<int> UpdateOneAsync(T viewModel, params string[] fields);
}
```

此处只定义了本项目中用到的新增和修改方法，删除将采用逻辑删除的方法（将记录的State值设为2，即删除状态）。

IRecordService接口继承IBaseService<RecordViewModel>接口，并添加了一个FindAsync方法，用于查询Record：

```
    public interface IRecordService: IBaseService<RecordViewModel>
    {
        Task<List<RecordViewModel>> FindAsync(RecordSearchModel SearchModel);
    }
```

对应的实现RecordService的代码如下：

```
    public class RecordService : IRecordService
    {
    private readonly IRecordRepository _recordRepository;
    private readonly IMapper _mapper;
    public RecordService(IRecordRepository RecordRepository, IMapper mapper)
    {
        _recordRepository = RecordRepository;
        _mapper = mapper;
    }
    public async Task InsertOneAsync(RecordViewModel viewModel)
    {
        var model = _mapper.Map<Record>(viewModel);
        model.CreateTime = DateTime.Now;
        await _recordRepository.InsertOneAsync(model);
    }
    public async Task<List<RecordViewModel>> FindAsync(RecordSearchModel searchModel)
    {
        var list = await _recordRepository.FindAllAsync(searchModel);
        return  _mapper.Map<List<Record>, List<RecordViewModel>>(list);
    }
```

```
    public async Task<int> UpdateOneAsync(RecordViewModel viewModel, params string[]
fields)
    {
        var model = _mapper.Map<Record>(viewModel);
        return await _recordRepository.UpdateOneAsync(model, fields);
    }
}
```

此处用到了一个组件 AutoMapper，用于 Model 和 ViewModel 的直接转换，下文会介绍它的配置和使用方法。

21.2.5 Repository 接口及实现

分别定义 Repository 接口和实现的基类，代码如下：

```
    public interface IBaseRepository<T> where T : BaseModel
    {
        T FindById(string id);

        List<T> FindAll();
        Task<List<T>> FindAllAsync();

        void InsertOne(T info);
        Task InsertOneAsync(T info);

        int UpdateOne(T info, params string[] fields);
        Task<int> UpdateOneAsync(T info, params string[] fields);

        int ReplaceOne(T info);
        Task<int> ReplaceOneAsync(T info);
    }

    public class BaseRepository<T> : IBaseRepository<T> where T : BaseModel
    {
        protected MongoHelper _mongoHelper;
        public T FindById(string id)
        {
            var builder = Builders<T>.Filter;
            var filter = builder.Eq(m => m._id, id);
            return _mongoHelper.Find(filter);
        }

        public List<T> FindAll()
        {
            return _mongoHelper.FindList<T>();
        }
```

```
        public async Task<List<T>> FindAllAsync()
        {
            return await _mongoHelper.FindListAsync<T>();
        }

        public void InsertOne(T info)
        {
            _mongoHelper.InsertOne(info);
        }

        public async Task InsertOneAsync(T info)
        {
            await _mongoHelper.InsertOneAsync(info);
        }

        public int UpdateOne(T info, params string[] fields)
        {
            return _mongoHelper.UpdateOne(info, fields: fields);
        }
        public async Task<int> UpdateOneAsync(T info, params string[] fields)
        {
            return await _mongoHelper.UpdateOneAsync(info, fields: fields);
        }

        public int ReplaceOne(T info)
        {
            return _mongoHelper.ReplaceOne(info);
        }

        public async Task<int> ReplaceOneAsync(T info)
        {
            return await _mongoHelper.ReplaceOneAsync(info);
        }
    }
```

此处用到了上文定义的 MongoHelper。这两个类可以完成大部分工作，所以 RecordRepository 类继承 BaseRepository 后只自定义添加了一个方法：

```
    public interface IRecordRepository : IBaseRepository<Record>
    {
        Task<List<Record>> FindAllAsync(RecordSearchModel searchModel);
    }
    public class RecordRepository : BaseRepository<Record>, IRecordRepository
    {
        public RecordRepository(MongoHelper mongoHelper)
        {
```

```
            _mongoHelper = mongoHelper;
        }
        public async Task<List<Record>> FindAllAsync(RecordSearchModel searchModel)
        {
            var builder = Builders<Record>.Filter;
            var filter = builder.Eq(m => m.State, searchModel.State);
            var list = await _mongoHelper.FindListAsync<Record>(filter);
            searchModel.RecordCount = list.Count;
            if (searchModel.IsPagination)
            {
                return list.OrderByDescending(m => m.CreateTime).Skip((searchModel.
PageIndex - 1) * searchModel.PageSize).Take(searchModel.PageSize).ToList();
            }
            return list.OrderByDescending(m => m.CreateTime).ToList();
        }
    }
```

21.2.6 Controller 与 Action

RecordController 的代码如下：

```
    [Route("api/[controller]")]
    public class RecordController : BaseController
    {
        private readonly IRecordService _recordService;
        public RecordController(IRecordService recordService)
        {
            _recordService = recordService;
        }

        [HttpGet]
        public async Task<ApiResult> Get(RecordSearchModel searchViewModel)
        {
            try
            {
                var list = await _recordService.FindAsync(searchViewModel);
                return new ApiResult<PagesModel<RecordViewModel>>(new Model.PagesModel
<RecordViewModel> (list,searchViewModel));
            }
            catch (Exception)
            {
                return new ApiResult(ReturnCode.GeneralError);
            }
        }

        [HttpPost]
        public async Task<ApiResult> Post([FromBody]RecordViewModel record)
```

```
        {
            if (record == null)
            {
                return new ApiResult(ReturnCode.ArgsError);
            }
            try
            {
                if (string.IsNullOrEmpty(record._id) && record.State == 1)
                {
                    record.CreateTime = DateTime.Now;
                    await _recordService.InsertOneAsync(record);
                }
                else if (!string.IsNullOrEmpty(record._id) && record.State == 2)
                {
                    await _recordService.UpdateOneAsync(record,"State");
                }
            }
            catch (Exception)
            {
                return new ApiResult(ReturnCode.GeneralError);
            }
            return new ApiResult(ReturnCode.Success);
        }
}
```

Get 方法用于查询身高和体重记录。Post 方法会根据记录的主键和 State 状态对记录进行新增和修改。可以看到，RecordController 继承自 BaseController。BaseController 是自定义的一个 Controller 的基类：

```
[Authorize]
public class BaseController : Controller
{
}
```

目前只是添加了[Authorize]属性，所有继承此类的 Controller 都要通过授权才能访问。对于例外的 Controller 或 Action（例如用于用户登录的 Action），可以通过[AllowAnonymous]属性取消授权要求。

21.2.7　AutoMapper 的使用

AutoMapper 是基于对象到对象约定的映射工具，例如本项目中将 Model 和 ViewModel 相互转换。在 NuGet 中搜索并安装它，如图 21-9 所示。

▲图 21-9

使用前需要在 Startup 中注册：

```
public void ConfigureServices(IServiceCollection services)
{
     services.AddAutoMapper(typeof(AutoMapperProfile));
}
```

参数 AutoMapperProfile 是一个配置文件，Model 之间的转换关系需要在此设置：

```
public class AutoMapperProfile : Profile
{
    public AutoMapperProfile()
    {
        CreateMap<UserViewModel, User>();
        CreateMap<User, UserViewModel>();

        CreateMap<RecordViewModel, Record>();
        CreateMap<Record, RecordViewModel>();
    }
}
```

类名可以自定义，只需要继承 AutoMapper.Profile，并在构造方法中设置映射关系即可。至此 API 项目的功能设置基本完成。

21.3 应用 JWT 进行用户认证

传统的 Web 应用一般采用 Cookies+Session 进行认证。但对于目前越来越多的 App、小程序来说，它们对应的服务端一般是 RESTful 类型的无状态的 API，再采用这样的认证方式不是很方便。而 JWT 这种无状态的、分布式的身份验证方式恰好符合这样的需求。

JWT（Json Web Token）基于开放标准（RFC 7519），是一种无状态的、分布式的身份验证方式，主要用于在网络应用环境间安全地传递声明。JWT 基于 JSON，所以它也像 JSON 一样可以在.NET、Java、JavaScript 和 PHP 等语言中使用。

21.3.1　JWT 的组成

JWT 就是下面这样一段字符串：


```
eyJhbGciOiJIUzI1NiIsInR5cCI6IkpXVCJ9.eyJodHRwOi8vc2NoZW1hcy54bWxzb2FwLm9yZy93cy8yMDA1
LzA1L2lkZW50aXR5L2NsYWltcy9uYW1laWRlbnRpZmllciI6IjAwMiIsImh0dHA6Ly9zY2hlbWFzLnhtbHNv
YXAub3JnL3dzLzIwMDUvMDUvaWRlbnRpdHkvY2xhaW1zL25hbWUiOiLmnY7lm5siLCJuYmYiOjE1NjU5Mj
MxMjIsImV4cCI6MTU2NTkyMzI0MiwiaXNzIjoiaHR0cDovL2xvY2FsaG9zdDo1NDIxNCIsImF1ZCI6Imh0dHA
6Ly9sb2NhbGhvc3Q6NTQyMTUifQ.Mrta7nftmfXeo_igBVd4rl2keMmm0rg0WkqRXoVAeik
```


它由 3 段“乱码”字符串通过两个“.”连接在一起组成，其官网提供了验证方式。JWT 的 3 个字符串分别对应图 21-10 右侧的 Header、Payload 和 VERIFY Signature 三部分。

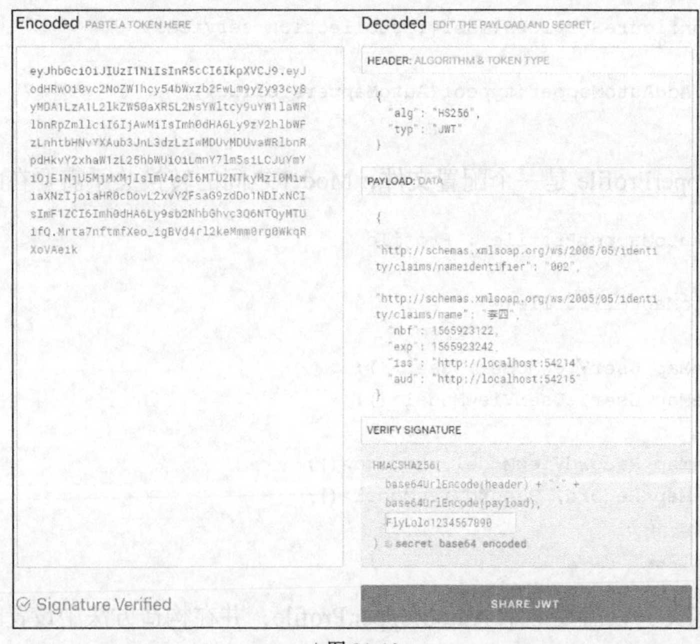

▲图 21-10

（1）Header。

Header 相关代码如下：

```
Header:
{
"alg": "HS256",
"typ": "JWT"
}
```

标识加密密钥为 HS256，Token 类型为 JWT，这部分通过 Base64Url 编码形成第一个字符串。

（2）Payload。

用于 JWT 信息存储，包含了多种声明（claims），它可以自定义多个声明添加到 PAYLOAD 中，系统也提供了一些默认的声明。

- **iss（issuer）**：签发人。
- **exp（expiration time）**：过期时间。
- **sub（subject）**：主题。
- **aud（audience）**：受众。
- **nbf（Not Before）**：生效时间。
- **iat（Issued At）**：签发时间。
- **jti（JWT ID）**：编号。

这部分通过 Base64Url 编码生成第二个字符串。

（3）Signature。

用于 Token 的验证。它的值类似表达式：

Signature = HMACSHA256 (base64UrlEncode (header) + "." + base64UrlEncode(payload), secret)

也就是说，它是通过将前两个字符串加密后生成的一个新字符串实现的。所以只有拥有同样密钥的人，才能通过前两个字符串获得相同的字符串，从而保证 Token 的真实性。

21.3.2 认证流程

认证流程如图 21-11 所示。

- **认证服务器**：用于用户的登录验证和 Token 的发放。
- **应用服务器**：业务数据接口，被保护的 API。
- **客户端**：一般为 App、小程序等。

认证流程如下。

（1）用户首先通过客户端登录，到认证服务器获取一个 Token。

（2）在访问应用服务器的 API 时，将获取的 Token 放置在请求的 Header 中。

（3）应用服务器验证该 Token，通过后返回对应的结果。

说明：这只是示例方案，实际项目中可能有所不同。

- 对于小型项目，认证服务和应用服务可能放在一起处理。本例通过分开的方式来实现，以便我们了解二者之间的认证流程。
- 对于复杂一些的项目，可能存在多个应用服务，用户获取的 Token 可以在多个分布式服务中被认证，这也是 JWT 的优势之一。

下面通过实际项目来看认证在 ASP.NET Core 中的应用。

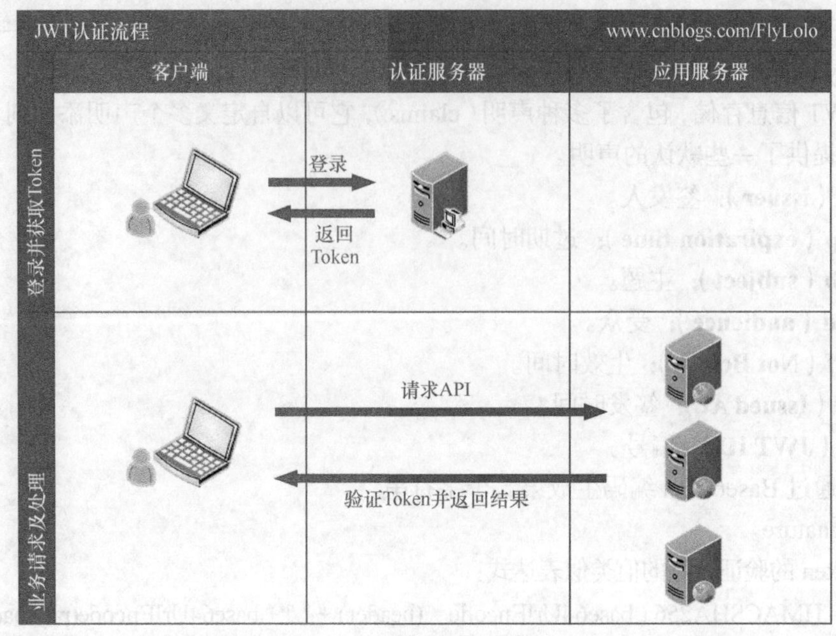

▲图 21-11

21.3.3　用户登录与 Token 的发放

图 21-11 描述了 Token 的认证流程。用户登录对应本项目中的 AccountController。

```
[Route("api/[controller]")]
public class AccountController : BaseController
{
    private readonly WXOptions _options;
    private readonly IUserService _userService;
    private readonly ITokenHelper _tokenHelper = null;
    public AccountController(IOptionsMonitor<WXOptions> options, IUserService
userService, ITokenHelper tokenHelper)
    {
        _options = options.Get("WXOptions");
        _userService = userService;
        _tokenHelper = tokenHelper;
    }

    [HttpGet]
    [AllowAnonymous]
    public JsonResult Get([FromServices] IHttpClientFactory httpClientFactory)
    {
        string loginCode;
        if (Request.Query.Keys.Contains("loginCode"))
        {
```

```
            loginCode = Request.Query["loginCode"];

            if (string.IsNullOrEmpty(loginCode))
            {
                return new JsonResult(new ApiResult(ReturnCode.ArgsError));
            }
        }
        else
        {
            return new JsonResult(new ApiResult(ReturnCode.ArgsError));
        }

        Code2Session session = null;
        string url = string.Format(_options.Code2Session, _options.AppId, _options.
Secret, loginCode);

        using (var client = httpClientFactory.CreateClient())
        {
            using var res = client.GetAsync(url);
            if (res.Result.StatusCode == System.Net.HttpStatusCode.OK)
            {
                var str = res.Result.Content.ReadAsStringAsync().Result;
                session = JsonConvert.DeserializeObject<Code2Session>(str);
            }
        }

        if (string.IsNullOrEmpty(session.Openid))
        {
            return new JsonResult(new ApiResult(ReturnCode.GeneralError));
        }
        //小程序返回的 OpenID 验证
        UserSearchModel userSearchModel = new UserSearchModel { WxOpenId = session.
Openid };
        UserViewModel user = _userService.Find(userSearchModel);

        if (null == user)
        {
            //用户名密码方式验证
            if (Request.Query.Keys.Contains("usercode"))
            {
                userSearchModel.UserCode = Request.Query["usercode"];
                user = _userService.Find(userSearchModel);
                if (null == user || (!user.Password.Equals(Request.Query["userPassword"])))
                {
                    return new JsonResult(new ApiResult(ReturnCode.LoginError));
                }
```

```
                user.NickName = Request.Query["nickName"];
                user.Gender = Request.Query["gender"];
                user.Country = Request.Query["country"];
                user.Province = Request.Query["province"];
                user.City = Request.Query["city"];
                user.Language = Request.Query["language"];
                user.AvatarUrl = Request.Query["avatarUrl"];
                user.WxOpenId = session.Openid;

                try
                {
                    _userService.UpdateOneAsync(user, "NickName", "Gender", "Country",
"Province", "City", "Language", "AvatarUrl", "WxOpenId");
                }
                catch (System.Exception)
                {
                    return new JsonResult(new ApiResult(ReturnCode.GeneralError));
                }
            }
            else
            {
                return new JsonResult(new ApiResult(ReturnCode.LoginError));
            }
        }
        var token = _tokenHelper.CreateToken(user);
        if (null == token) return StatusCode(401);
        return new JsonResult(new ApiResult<ComplexToken>(token, ReturnCode.Success));
    }
}
```

代码中有一行注释为“用户名密码方式验证”。以这行注释为界，将 Action 分为两部分。

下半部分的逻辑是根据用户提交的登录信息中的 UserCode 查找对应的用户，若找到对应的用户，则验证密码正确（示例项目中密码未做加密处理）。若验证成功，则通过_tokenHelper 生成 Token 作为请求结果返回。这是普通项目中常见的逻辑。

上半部分的代码则是针对小程序登录的，目的是避免每次打开小程序都需要输入用户名和密码。当第一次打开小程序时，采用用户名密码方式验证登录，登录成功后保存登录用户的 OpenID（用户针对本小程序的唯一 ID）。当第二次打开小程序时，首先验证请求中是否包含了用户的 OpenID。若是，则根据该 OpenID 去查找对应的用户并自动完成登录功能，跳过用户名和密码的验证过程。

总结：第一次登录实际是经历了小程序登录和用户名密码登录两次登录验证，验证通过后，将小程序提供的 OpenID 与用户信息绑定，第二次则是自动完成小程序的登录功能，跳过用户名和密码验证过程。

小程序官方的开发文档中针对用户的登录流程描述如图 21-12 所示。

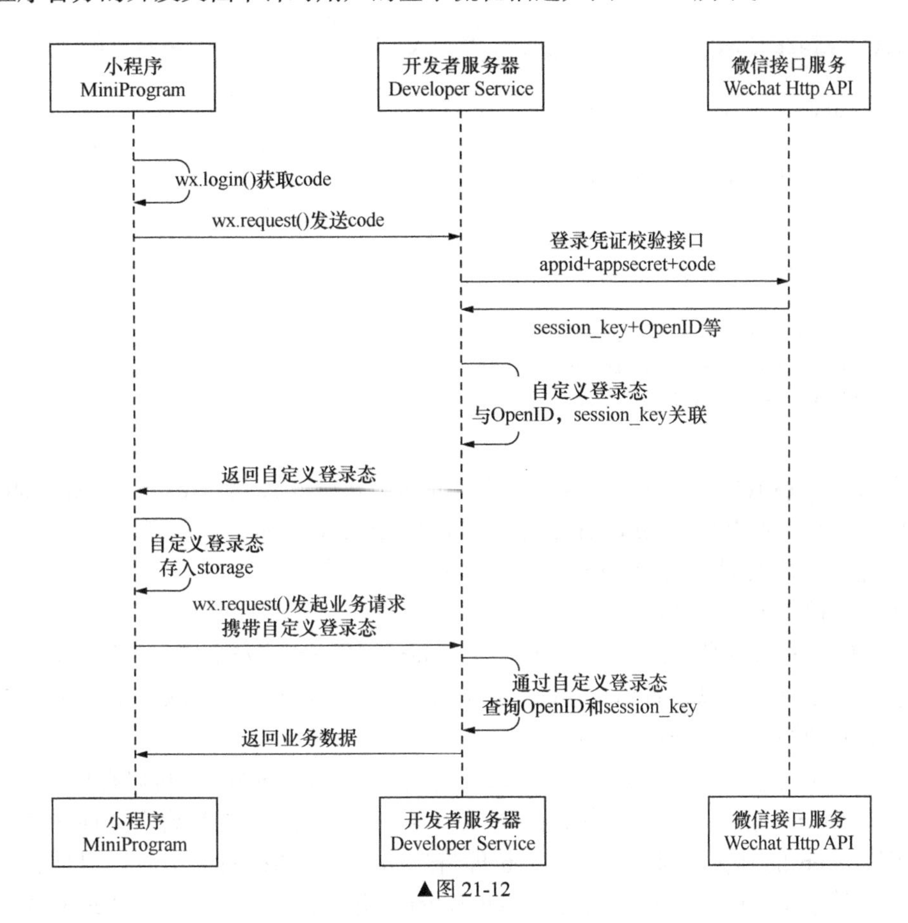

▲图 21-12

说明如下。

- 首先调用 wx.login() 获取临时登录凭证（code），并回传到开发者服务器。
- 调用 auth.code2Session 接口，换取用户唯一标识（OpenID）和会话密钥（session_key）。
- 之后开发者服务器可以根据用户标识来生成自定义登录态，用于后续业务逻辑中前后端交互时识别用户身份。

注意：

- 会话密钥是对用户数据进行“加密签名”的密钥。为了实现自身的数据安全，开发者服务器不应该把会话密钥下发到小程序，也不应该对外提供这个密钥。
- 临时登录凭证只能使用一次。

图 21-12 和对应的说明引用自官方提供的开发文档，描述了用户的认证过程。说明中“开发者服务器可以根据用户标识来生成自定义登录态”以及图 21-12 中的“返回自定义登录态”对于本项目来说就是 Token 的发放。

下面来看 Token 的生成代码：

```
public class Token
{
    public string TokenContent { get; set; }

    public DateTime Expires { get; set; }
}

public class ComplexToken
{

    public Token AccessToken { get; set; }
    public Token RefreshToken { get; set; }
    public UserViewModel User {get;set;}
}
```

返回的 ComplexToken 实际上包含了两个 Token，分别是 AccessToken 和 RefreshToken。AccessToken 用于认证请求，RefreshToken 用于刷新 AccessToken。

在读到上面的内容时，读者可能会有如下一些疑问。

（1）Token 被盗了怎么办？

在启用 HTTPS 的情况下，Token 被放在 Header 中还是比较安全的。另外，Token 的有效期不要设置过长，例如可以设置为 1 小时（微信公众号的网页开发的 Token 有效期为 2 小时）。

（2）Token 到期了如何处理？

理论上，Token 过期程序应该是返回登录页面，但这样太不友好了。可以在后台根据 Token 的过期时间定期请求新的 Token，所以希望能在 Token 过期之前，在后台自动地对 Token 进行"续期"，即 Token 的刷新。当 AccessToken 即将过期时，例如提前 5 分钟，可使客户端利用 RefreshToken 请求指定的 API 获取新的 AccessToken，并更新本地存储中的 AccessToken。

（3）为什么不直接用 RefreshToken 去认证，还让 AccessToken 一次次地被刷新？

AccessToken 用于每次业务请求，而 RefreshToken 只会在调用刷新 AccessToken 的接口时才会使用，不会使用得太频繁，所以被盗的概率相对较小。并且 AccessToken 有效期较短，例如 1 小时，这样安全一些。RefreshToken 有效期可以设置得长一些，例如一天、一周等。

TokenHelper 的代码如下：

```
public enum TokenType
{
    AccessToken = 1,
    RefreshToken = 2
}
public class TokenHelper : ITokenHelper
{
    private readonly IOptions<JWTConfig> _options;
    public TokenHelper(IOptions<JWTConfig> options)
```

```
    {
        _options = options;
    }

    public Token CreateAccessToken(UserViewModel user)
    {
        Claim[] claims = { new Claim(ClaimTypes.NameIdentifier, user.UserCode), new
Claim(ClaimTypes.Name, user.UserName) };

        return CreateToken(claims, TokenType.AccessToken);
    }

    public ComplexToken CreateToken(UserViewModel user)
    {
        Claim[] claims = { new Claim(ClaimTypes.NameIdentifier, user.UserCode), new
Claim(ClaimTypes.Name, user.UserName) };

        return new ComplexToken { AccessToken = CreateToken(claims, TokenType.
AccessToken),
            RefreshToken = CreateToken(claims, TokenType.RefreshToken),
            User = user
        };
    }

    // <summary>
    // 用于创建 AccessToken 和 RefreshToken
    // 这里 AccessToken 和 RefreshToken 只是过期时间不同，"实际项目" 中二者的 claims 内容可能会不同
    // 因为 RefreshToken 只是用于刷新 AccessToken，其内容可以简单一些
    // 而 AccessToken 可能会附加一些其他的 claim
    // </summary>
    // <param name="claims"></param>
    // <param name="tokenType"></param>
    // <returns></returns>
    private Token CreateToken(Claim[] claims, TokenType tokenType)
    {
        claims = claims.Append(new Claim("TokenType", tokenType.ToString())).ToArray();
        var now = DateTime.Now;
        var expires = now.Add(TimeSpan.FromMinutes(tokenType.Equals(TokenType.
AccessToken) ? _options.Value.AccessTokenExpiresMinutes : _options.Value.RefreshToken
ExpiresMinutes));
        var token = new JwtSecurityToken(
            issuer: _options.Value.Issuer,
            audience: _options.Value.Audience,
            claims: claims,
            notBefore: now,
            expires: expires,
```

```
            signingCredentials: new SigningCredentials(new SymmetricSecurityKey
(Encoding.UTF8.GetBytes(_options.Value.IssuerSigningKey)), SecurityAlgorithms.HmacSha256));
        return new Token { TokenContent = new JwtSecurityTokenHandler().WriteToken
(token), Expires = expires };
    }

    public Token RefreshToken(ClaimsPrincipal claimsPrincipal)
    {
        if (claimsPrincipal.Claims.Any(m=>m.Type.Equals("TokenType") && m.Value.Equals
(TokenType.RefreshToken.ToString())))
        {
            var code = claimsPrincipal.Claims.FirstOrDefault(m => m.Type.Equals
(ClaimTypes.NameIdentifier));
            var name = claimsPrincipal.Claims.FirstOrDefault(m => m.Type.Equals
(ClaimTypes.Name));
            if (null != code)
            {
                return CreateAccessToken(new UserViewModel { UserCode = code.Value,
UserName = name.Value });
            }
        }

        return null;
    }
}
```

这里主要包括 Token 的发放和刷新两部分。在 CreateToken 方法中通过 JwtSecurityToken 方法创建 Token，关键参数如下。

- **Issuer**：Token 发布者。
- **Audience**：Token 接收者。
- **expires**：过期时间。
- **IssuerSigningKey**：签名密钥。

根据 Token 类型的不同设置了不同的过期时间，并把 Token 的类型 TokenType 作为一个 claim 加入 Token，目的是在请求认证时只能使用 AccessToken，在刷新 Token 时只能使用 RefreshToken，因此在 RefreshToken 方法中添加了判断语句。

除此之外，还需要创建一个用于刷新 Token 的 Controller：

```
[Route("Token")]
[Authorize]
public class TokenController : Controller
{
    private readonly ITokenHelper tokenHelper;
    public TokenController(ITokenHelper _tokenHelper)
    {
```

```
        tokenHelper = _tokenHelper;
    }

    [HttpGet]
    public ApiResult Get()
    {
        return new ApiResult<Token>(tokenHelper.RefreshToken(Request.HttpContext.User),
ReturnCode.Success);
    }
}
```

有了认证方法，现在需要使它生效，在 Startup 中添加相关注册：

```
public class Startup
{
    public void ConfigureServices(IServiceCollection services)
    {
        services.AddControllers();
        services.AddHttpClient();

        // 省略部分代码
        #region //读取配置信息
        services.AddSingleton<ITokenHelper, TokenHelper>();
        services.Configure<JWTConfig>(Configuration.GetSection("JWT"));
        JWTConfig config = new JWTConfig();
        Configuration.GetSection("JWT").Bind(config);
        #endregion


        #region //启用 JWT 认证
        services.AddAuthentication(options =>
        {
            options.DefaultAuthenticateScheme = JwtBearerDefaults.AuthenticationScheme;
            options.DefaultChallengeScheme = JwtBearerDefaults.AuthenticationScheme;
        }).
        AddJwtBearer(options =>
        {
            options.TokenValidationParameters = new TokenValidationParameters
            {
                ValidIssuer = config.Issuer,
                ValidAudience = config.Audience,
                IssuerSigningKey = new SymmetricSecurityKey(Encoding.UTF8.GetBytes
(config.IssuerSigningKey)),
                //ClockSkew = TimeSpan.FromMinutes(5)
            };
        });
        #endregion
```

```
    }

    // This method gets called by the runtime. Use this method to configure the HTTP
request pipeline.
    public void Configure(IApplicationBuilder app, IWebHostEnvironment env)
    {
        if (env.IsDevelopment())
        {
            app.UseDeveloperExceptionPage();
        }

        app.UseRouting();
        app.UseAuthentication();
        app.UseAuthorization();

        app.UseEndpoints(endpoints =>
        {
            endpoints.MapControllers();
        });
    }
}
```

在 Configure 方法中通过 UseAuthentication 方法启用认证，在 ConfigureServices 方法中对认证进行配置，设置如下参数。

- **ValidIssuer**：用于和接收到的 Token 的发布者进行比对。
- **ValidAudience**：用于和接收到的 Token 的接收者进行比对。
- **ClockSkew**：缓冲过期时间，默认为 5 分钟，即 Token 过期 5 分钟内仍然是有效的。
- **IssuerSigningKey**：签名密钥。

通过 TokenValidationParameters 的构造方法查看参数的默认值，代码如下：

```
public TokenValidationParameters()
{
    RequireExpirationTime = true;
    RequireSignedTokens = true;
    SaveSigninToken = false;
    ValidateActor = false;
    ValidateAudience = true;
    ValidateIssuer = true;
    ValidateIssuerSigningKey = false;
    ValidateLifetime = true;
    ValidateTokenReplay = false;
}
```

可见 TokenValidationParameters 中还有一些其他参数可以设置，例如可以设置是否验证发布者、是否验证接收者等，这里设置了默认值，根据需求修改即可。

在 appsettings.json 中配置认证信息如下：

```
"JWT": {
    "Issuer": "FlyLolo",
    "Audience": "TestAudience",
    "IssuerSigningKey": "FlyLolo1234567890",
    "AccessTokenExpiresMinutes": "30"
 }
```

为了测试简单，可以暂时注释掉登录方法中关于微信小程序的处理，这样可以直接通过 Fiddler 或 Postman 发送用户名密码测试。登录成功会返回如下两个 Token：


```
{"accessToken":{"tokenContent":"eyJhbGciOiJIUzI1NiIsInR5cCI6IkpXVCJ9.eyJodHRwOi8vc2
NoZW1hcy54bWxzb2FwLm9yZy93cy8y
MDA1LzA1L2lkZW50aXR5L2NsYWltcy9uYW1laWRlbnRpZmllciI6IjAwMiIsImh0dHA6Ly9zY2hlbWFzLnhtb
HNvYXAub3JnL3dzLzIwMDUvMDUva
WRlbnRpdHkvY2xhaW1zL25hbWUiOiLmnY7lm5siLCJodHRwOi8vc2NoZW1hcy5taWNyb3NvZnQuY29tL3dzLz
IwMDgvMDYvaWRlbnRpdHkvY2xhaW
1zL3JvbGUiOlsiVGVzdFB1dEJvb2tSb2xlIiwiVGVzdFB1dFN0dWRlbnRSb2xlIl0sIm5iZiI6MTU2NjgwNjQ
3OSwiZXhwIjoxNTY2ODA4Mjc5LCJ
pc3MiOiJGbHlMb2xvIiwiYXVkIjoiVGVzdEF1ZGllbmNlIn0.wlMorS1V0xP0Fb2MDX7jI7zsgZbb2Do3u78B
AkIIwGg",
"expires":"2019-08-26T22:31:19.5312172+08:00"},

"refreshToken":{"tokenContent":"eyJhbGciOiJIUzI1NiIsInR5cCI6IkpXVCJ9.eyJodHRwOi8vc2No
ZW1hcy54bWxzb2FwLm9yZy93cy8y
MDA1LzA1L2lkZW50aXR5L2NsYWltcy9uYW1laWRlbnRpZmllciI6IjAwMiIsImh0dHA6Ly9zY2hlbWFzLnhtb
HNvYXAub3JnL3dzLzIwMDUvMDUva
WRlbnRpdHkvY2xhaW1zL25hbWUiOiLmnY7lm5siLCJodHRwOi8vc2NoZW1hcy5taWNyb3NvZnQuY29tL3dzLz
IwMDgvMDYvaWRlbnRpdHkvY2xhaW
1zL3JvbGUiOlsiVGVzdFB1dEJvb2tSb2xlIiwiVGVzdFB1dFN0dWRlbnRSb2xlIl0sIm5iZiI6MTU2NjgwNjQ
3OSwiZXhwIjoxNTY3NDExMjc5LCJ
pc3MiOiJGbHlMb2xvIiwiYXVkIjoiUmVmcmVzaFRva2VuQXVkaWVuY2UifQ.3EDi6cQBqa39-ywq2EjFGiM8W
2KY5l9QAOWaIDi8FnI",
"expires":"2019-09-02T22:01:19.6143038+08:00"}}
```


配置好后，添加[Authorize]标识，使 Controller 或 Action 需要通过认证才能被访问：

```
[Route("api/[controller]")]
[Authorize]
public class RecordController : BaseController
```

访问需要认证的 Action 时，在 Header 中需添加 Authorization: bearer AccessToken。但存在一个问题：只要有了 Token，就可以访问所有 API 了，这显然不是我们想要的。我们希望通过不同的角色和权限设置，使用户只能访问自己有权限的 API，这就会涉及用户的授权。

21.4 自定义用户授权

认证（authentication）与授权（authorization）经常在一起工作，所以有时我们会分不清楚，这两个英文单词也类似。举例来说，我刷门禁卡进入公司，门禁“认证”了我是这里的员工，可以进入，但进入公司以后，并不是所有房间我都可以进，我能进入哪些房间，需要公司的“授权”。这就是认证和授权的区别。

ASP.NET Core 提倡的是基于声明（Claim）的授权，关于这个声明的代码如下：

```
Claim[] claims = new Claim[] { new Claim(ClaimTypes.NameIdentifier, user.Code), new
Claim(ClaimTypes.Name, user.Name) };
```

这是一个声明的集合，它包含两个声明，用于保存用户的唯一 ID 和用户名，我们还可以添加更多的声明。对应 Claim，还有 ClaimsIdentity 和 ClaimsPrincipal 两个类型。

ClaimsIdentity 相当于一个证件，例如前面提及的门禁卡；ClaimsPrincipal 则是证件的持有者，也就是“我本人”。那么对应的 Claim 就是门禁卡内存储的一些信息，例如证件号、持有人姓名等。

除了门禁卡，我还有身份证、银行卡等，也就是说一个 ClaimsPrincipal 中可以有多个 ClaimsIdentity，而一个 ClaimsIdentity 中可以有多个 Claim。ASP.NET Core 的授权模型大概就是这样一个体系。

ASP.NET Core 提供了多种授权方式。有基于“角色”的授权，即给用户分配几种角色，然后给对应的 Controller 或 Action 设置只允许拥有相应角色的用户访问；还有基于声明的授权，即用户拥有并符合某种声明才可以访问。这两种方式都比较简单，适合一些角色比较单一的小项目。复杂项目经常存在这样的需求：一个用户可以有多个角色，每个角色对应多个可访问的 API（将授权细化到具体的 Action），用户还可以被特殊地授予某个 API 的权限。

采用上面的两种方式难以实现复杂的需求，好在 ASP.NET Core 提供了方便的扩展方式。

21.4.1　样例数据

将上面的需求汇总，最终形成如下形式的代码：

```
public static class TemporaryData
{
    public readonly static List<UserPermissions> UserPermissions = new List
<UserPermissions> {
        new UserPermissions {
            UserCode = "001",
            Roles = new List<Role> {
                new Role { Code = "System.Manager", Name = "系统管理"},
                new Role { Code = "Record.Viewer", Name = "记录浏览" },
                new Role { Code = "Record.Operator", Name = "记录操作"}
```

```
            },
            Permissions = new List<Permission> {
                new Permission { Code = "User.Write", Name = "用户管理", Controller = "user",
Action="post"},
                new Permission { Code = "Record.Read", Name = "记录读取", Controller =
"record",Action="get" },
                new Permission { Code = "Record.Write", Name = "记录修改", Controller =
"record",Action="post"}
            }
        },
        new UserPermissions {
            UserCode = "002",
            Roles = new List<Role> {
                new Role { Code = "Record.Viewer", Name = "记录浏览" }
            },
            Permissions = new List<Permission> {
                new Permission { Code = "Record.Read", Name = "记录读取", Controller =
"record",Action="get" }
            }
        },
    };

    public static UserPermissions GetUserPermission(string userCode)
    {
        return UserPermissions.FirstOrDefault(m => m.UserCode.Equals(userCode));
    }
}
```

此处为虚拟数据，模拟从数据库或缓存中读取用户相关的权限。在实际项目中，根据项目的大小可能会采用不同力度的权限管理方式。涉及的类如下：

```
public class UserViewModel:BaseViewModel
{
    public string UserCode { get; set; }
    public string UserName { get; set; }
    public string WxOpenId { get; set; }
    public string NickName { get; set; }
    public string AvatarUrl { get; set; }
    public string Gender { get; set; }
    public string Country { get; set; }
    public string Province { get; set; }
    public string City { get; set; }
    public string Language { get; set; }
    public string Telephone { get; set; }
    public string Password { get; set; }
    public int State { get; set; }
```

```
    public UserPermissions UserPermissions { get; set; }
}

public class UserPermissions
{
    public string UserCode { get; set; }
    public List<Role> Roles { get; set; }
    public List<Permission> Permissions { get; set; }
}

public class Permission
{
    public string Code { get; set; }
    public string Name { get; set; }
    public string Controller { get; set; }

    public string Action { get; set; }
}

public class Role
{
    public string Code { get; set; }
    public string Name { get; set; }
}
```

21.4.2　自定义授权处理

下面根据样例数据来制定相应的处理程序。这里涉及 IAuthorizationRequirement 和 AuthorizationHandler 两个内容。

IAuthorizationRequirement 是一个空的接口，主要提供授权所需要满足的“要求”或“规则”。AuthorizationHandler 对请求和“要求”进行联合处理。

新建一个 PermissionRequirement 实现 IAuthorizationRequirement 接口：

```
public class PermissionRequirement: IAuthorizationRequirement
{
    public List<UserPermissions> UsePermissionList { get { return TemporaryData.
UserPermissions; } }
}
```

它的“要求”就是用户的权限列表，用户的权限列表中包含当前访问的 API，授权则通过，否则不通过。

判断语句放在新建的 PermissionHandler 中：

```
public class PermissionHandler : AuthorizationHandler<PermissionRequirement>
{
```

```
    protected override Task HandleRequirementAsync(AuthorizationHandlerContext context,
PermissionRequirement requirement)
    {
        if (context.User.Identity.IsAuthenticated && context.User.Claims.Any(m => m.
Type.Equals("TokenType") && m.Value.Equals(TokenType.AccessToken.ToString())))
        {
            var code = context.User.Claims.FirstOrDefault(m => m.Type.Equals
(ClaimTypes.NameIdentifier));
            if (null != code)
            {
                UserPermissions userPermissions = requirement.UsePermissionList.
FirstOrDefault(m => m.UserCode.Equals(code.Value.ToString()));
                var routContext = (context.Resource as Microsoft.AspNetCore.Routing.
RouteEndpoint);

                if (null != userPermissions && userPermissions.Permissions.Any(m =>
m.Controller.ToLower().Equals(routContext.RoutePattern.Defaults["Controller"].
ToString().ToLower()) && m.Action.ToLower().Equals(routContext.RoutePattern.
Defaults["Action"].ToString().ToLower())))
                {
                    context.Succeed(requirement);
                }
                else
                {
                    context.Fail();
                }
            }
            else
            {
                context.Fail();
            }
        }
        else
        {
            context.Fail();
        }

        return Task.CompletedTask;
    }
}
```

逻辑是判断当前请求的 Controller 和 Action 是否在当前用户的权限列表中。

最后需要在 Startup 的 ConfigureServices 中添加授权注册：

```
services.AddAuthorization(options => options.AddPolicy("Permission", policy =>
policy.Requirements.Add(new PermissionRequirement())));
```

```
services.AddSingleton<IAuthorizationHandler, PermissionHandler>();
```

在 Startup 的 Configure 方法中启用授权：

```
app.UseAuthorization();
```

为了方便，对 BaseController 进行认证和授权设置，代码如下：

```
[Authorize(Policy = "Permission")]
public class BaseController : Controller
{
}
```

这样所有继承此基类的 Controller 都会自动进行认证和授权的验证。

21.5 使用 Swagger 生成 Web API 的帮助页

API 就是给别人用的，因此要提供接口文档。当 API 更新之后还要更新该文档，非常麻烦。Swagger 可以自动完成这个工作，让我们更加专注于业务和代码，它不但能自动生成接口信息，甚至连 UI 都做好了。

在 NuGet 中下载安装相关的包。搜索 Swagger，建议下载 5.0 版本以上，因为写作本书时还没有出 5.0 正式版，所以勾选“包含预发行版”选项，安装目前最新的 v5.0.0-rc4 版本，单击“安装”按钮，同时会安装相关的包，如图 21-13 所示。

▲图 21-13

在 Startup 中添加并配置 Swagger 中间件。首先将 Swagger 生成器添加到 Startup.ConfigureServices 方法中注册：

```
public void ConfigureServices(IServiceCollection services)
{
    services.AddSwaggerGen(c =>
    {
```

```
        c.SwaggerDoc("v1", new OpenApiInfo { Title = "My API", Version = "v1" });
    });
}
```

然后添加相应的中间件，放在 UseRouting 前：

```
public void Configure(IApplicationBuilder app)
{
    // 启用中间件
    app.UseSwagger();

    // 启用 UI 显示
    app.UseSwaggerUI(c =>
    {
        c.SwaggerEndpoint("/swagger/v1/swagger.json", "My API V1");
    });

app.UseRouting();

//省略部分代码
}
```

最后运行项目并访问地址 http://localhost:63026/swagger，就可以看到如图 21-14 所示的“接口文档”了。

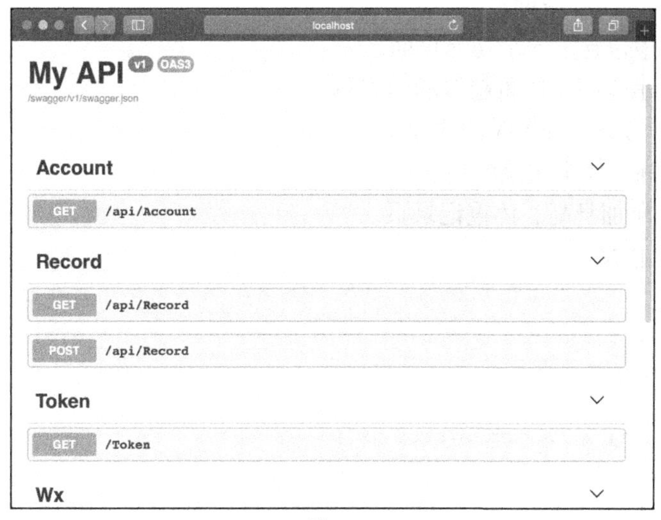

▲图 21-14

21.6 微信小程序

打开微信开发者工具，新建微信小程序项目，最终项目结构如图 21-15 所示。

```
▸ ec-canvas
▸ images
▾ pages
  ▸ charts
  ▸ launch
  ▾ note
    ▸ index
    ▸ noteAdd
  ▸ user
▾ style
    weui.wxss
▸ utils
  app.js
  app.json
  app.wxss
  project.config.json
  sitemap.json
```

▲图 21-15

- ec-canvas：引用的 echarts 图表组件。
- images：图片目录。
- style：微信官方的样式库。
- utils：封装的一些帮助类。
- pages：其中包含了 5 个功能页面。
- charts：显示身高、体重趋势的图表页。
- launch：刚打开小程序时的欢迎页。
- index：身高、体重记录的列表页。
- noteAdd：添加身高、体重记录页。
- user：个人信息页。

21.6.1　欢迎页

launch.js 的代码如下：

```
var http = require('../../utils/http.js')
var token = require('../../utils/token.js')
//获取应用实例
const app = getApp()
var that
Page({
  data: {
    userInfo: app.globalData.userInfo,
    loginState: 0,
```

```
    canIUse: wx.canIUse('button.open-type.getUserInfo'),
    message: ""
  },
  onLoad: function() {
    var state = 0;
    that = this;
    // 获取用户信息
    wx.getSetting({
      success: res => {
        //是否已授权
        if (res.authSetting['scope.userInfo']) {
          if (that.data.userInfo == null) {
            wx.getUserInfo({
              success: function(res) {
                app.globalData.userInfo = res.userInfo;
                that.setData({
                  userInfo : res.userInfo
                });
              }
            });
          }
          state = this.userLogin({});
        } else {
          that.setState(-1, "");
        }
      }
    })
  },
  userLogin: function(userInfo) {
    var state = 1;
    // 已授权后再登录，验证用户信息
    wx.login({
      success: res => {
        userInfo.loginCode = res.code;
        // 发送 res.code 到后台换取 OpenID、SessionKey、UnionID
        http.httpGet(
          "/api/Account",
          userInfo,
          function(res) {
            if (res.statusCode == 200 && res.data.code == 0) {
              token.setToken(res.data.data);
              that.toList();
            } else {
              that.setState(-8, " ");
            }
          },
```

```
        function(res) {
          that.setState(-8, " "); //temp  -16
        }, false
      );
    }
  })
},
getUserInfo: function(e) {
  app.globalData.userInfo = e.detail.userInfo;
  that.setData({
    userInfo: e.detail.userInfo,
  })
  that.userLogin(e.detail.userInfo);
},
setState: function(state, msg) {
  //console.log("loginState: " + state);
  app.globalData.loginState = state;
  if (!msg || msg.length == 0) {
    //0: 无状态（默认）1: 已授权  -1: 未授权  2: 待审核  -2: 审核未通过  8: 已登录  -8: 登录失败 -16 连接服务器失败
    switch (state) {
      case 2:
        msg = "已提交申请，请等待管理员审核。";
        break;
      case 3:
        msg = "未通过管理员审核。";
        break;
      case -8:
        msg = "登录失败。";
        break;
      case -16:
        msg = "连接服务器失败。";
        break;
      default:
        break;
    };
  }
  this.setData({
    loginState: state,
    message: msg
  });
},
toList: function() {
  wx.switchTab({
    url: "/pages/note/index/index"
  })
```

```
  },
  formSubmit: function(e) {
    try{
      var userInfo = that.data.userInfo;
      console.log("submit", userInfo);
      Object.assign(userInfo, e.detail.value);
      that.userLogin(userInfo);
    }
    catch(e){
      wx.showToast(e);
    }

  }
})
```

在执行 onLoad 时判断用户是否已授权，若未授权，则显示授权按钮，要求用户授权；若已授权，则进入 userLogin 方法进行用户登录。通过 wx.login 方法获取登录用的 code，并传给服务端的 AccountController 的 Get 方法。登录成功后会获取两个 Token 和用户信息，调用 token.setToken(res.data.data)方法进行处理，该方法写在 utils 的 token.js 中，代码如下：

```
const app = getApp();

const setToken = function(data) {
  app.globalData.userInfo = data.user;
  app.globalData.accessToken = data.accessToken;
  app.globalData.refreshToken = data.refreshToken;
  refreshToke();
}

const refreshToke = function() {
  var datestr = app.globalData.accessToken.expires.replace('T',' ').substr(0,16);

  var times = Date.parse(datestr) - new Date().getTime() - 5 * 60 * 1000;

  setTimeout(getToken, times);
}
const getToken = function() {
  wx.request({
    url: app.globalData.server + "/token",
    data: null,
    header: {
      'Authorization': "bearer " + app.globalData.refreshToken.tokenContent
    },
    success: function(res) {
      if (res.statusCode == 200 && res.data.code == 0) {
        console.log("token", res);
```

```
        app.globalData.accessToken = res.data.data;
        refreshToke();
      }
    },
    fail: function(res) {
      wx.navigateTo({
        url: '/pages/launch/launch'
      })
    }
  })
}

module.exports = {
  setToken: setToken
}
```

setToken 方法首先会将用户信息和两个 Token 分别赋值给 app.globalData 的对应变量备用，然后调用 refreshToken 方法定时刷新 accessToken。原理是先判断此 accessToken 的到期时间距离当前时间还有多久，通过设置 setTimeout，在过期前 5 分钟执行刷新请求，请求成功后，将获取的新 accessToken 赋值给 app.globalData.accessToke，并在此判断新的 accessToken 的过期时间，设定定时器。如此循环执行，保证 accessToken 的有效性。

21.6.2　列表页

index.wxml 文件的页面主要是一个表格，表格的数据采用 scroll-view 滚动展示，设置了当滚动到底部时会触发 scrollToLower 方法，代码如下：

```
<view>
  <view class="kind-list-item">
    <view class="kind-list-item-hd">
      <view class="kind-list-text">记录列表</view>
      <image class="kind-list-img" src="../../../images/add2.png" bindtap="addRecord">
</image>
    </view>
  </view>

  <view class="grids" style='height:90%'>
    <view class="grid">
      <p class="grid__label">时 间</p>
    </view>
    <view class="grid">
      <p class="grid__label">类 型</p>
    </view>
    <view class="grid">
      <p class="grid__label">值</p>
```

```
    </view>
    <view class="grid">
      <p class="grid__label">操作</p>
</view>

    <scroll-view scroll-y="true" bindscrolltolower="scrollToLower" style="height:
{{boxHeight}}px">
      <!-- 列表 -->
      <view wx:for='{{list}}' wx:key="_id">
        <view class="grid">
          <p class="grid__label">{{m1.formatDate(item.createTime)}}</p>
        </view>
        <view class="grid">
          <p class="grid__label">{{item.recordType == 1?'身高':'体重'}}</p>
        </view>
        <view class="grid">
          <p class="grid__label">{{item.value}}</p>
        </view>
        <view class="grid">
          <button class="mini-btn" data-id="{{item._id}}"data-index="{{index}}"
type="warn" size="mini" bindtap="deleteRecord">删除</button>
        </view>

      </view>
    </scroll-view>
  </view>
</view>

<wxs module="m1">
  var formatDate = function(date) {
    date = date.length > 0 ? date.replace("T", " ").substring(2, 10) : "";
    return date;
  }
  module.exports.formatDate = formatDate;
</wxs>
```

对应的 index.js 代码如下：

```
var http = require('../../../utils/http.js')
var app = getApp()
var that

Page({
  /**
   * 页面的初始数据
   */
  data: {
```

```
    list: [],
    pageIndex:1,
    pageSize:20,
    recordCount: 0
  },

  /**
   * 生命周期函数——监听页面加载
   */
  onLoad: function (options) {
    that = this;
    let res = wx.getSystemInfoSync();
    let boxHeight = res.windowHeight - 135;
    this.setData({
      'boxHeight': boxHeight
    })
  },
  onShow:function()
  {
    that.setList();
  },
  onPullDownRefresh: function () {
    that.setList();
  },
  scrollToLower: function () {
    if (that.data.pageSize * that.data.pageIndex < that.data.recordCount){
      that.setList(1);
    }
  },
  setList:function(addPage = 0)
  {
    http.httpGet(
      "/api/record",
      { UserCode: app.globalData.userInfo.userCode, PageIndex: that.data.pageIndex +
addPage, PageSize: that.data.pageSize, IsPagination: true, State: 1 },
      function(res){
        if (res.statusCode == 200 && res.data.code == 0) {
          that.setData({
            list: that.data.list.concat(res.data.data.items) ,
            pageIndex: res.data.data.pageIndex,
            recordCount: res.data.data.recordCount
          });
          console.log(that.data.recordCount,that.data.list);
        }
      },
      function (res) {
```

```
        }
      );
    },
    addRecord:function()
    {
      wx.navigateTo({
        url: '/pages/note/noteAdd/noteAdd'
      })
    },
    deleteRecord: function (e) {

      http.httpPost(
        "/api/record",
        {
          _id: e.currentTarget.dataset.id,
          State: 2,
        },
        function (res) {

        }
      );
      that.data.list.splice(e.currentTarget.dataset.index, 1);
      that.setData({ list: that.data.list });
    }
})
```

默认获取第一页的内容，当scroll-view滚动到底部时，修改每页请求的总行数，获取下一页的内容。列表的删除按钮会调用deleteRecord方法，传递当前记录的ID和删除状态State=2给服务端的/api/record的Post请求，将当前记录标记为删除状态。

21.6.3 图表页

charts.wxml的内容如下：

```
<view class="page">
  <ec-canvas id="mychart" canvas-id="mychart-line" ec="{{ ec }}"></ec-canvas>
</view>
```

主要是引用echarts的图表标签，没有其他内容。

chart.js的代码如下：

```
import * as echarts from '../../ec-canvas/echarts';
var http = require('../../utils/http.js');
var util = require('../../utils/util.js');
const app = getApp();
var that
```

```
var Chart = null;
Page({
  data: {
    ec: {
      lazyLoad: true
    }
  },
  onLoad: function (options) {
    this.echartsComponnet = this.selectComponent('#mychart');
    that = this;
    this.setDataList();

  },
  setDataList: function () {
    http.httpGet(
      "/api/Record",
      { State: 1, IsPagination: false },
      function (res) {
        if (res.statusCode == 200 && res.data.code == 0) {
          console.log("table", res);
          that.setData({ tableRows: res.data.data.items });
          that.setChart();
        }
      }
    );
  },
  setChart: function () {
    //如果是第一次绘制
    if (!Chart) {
      Chart = this.init_echarts(); //初始化图表
    } else {
      this.setOption(Chart); //更新数据
    }
  },
  //初始化图表
  init_echarts: function () {
    this.echartsComponnet.init((canvas, width, height) => {
      // 初始化图表
      Chart = echarts.init(canvas, null, {
        width: width,
        height: height
      });
      // Chart.setOption(this.getOption());
      this.setOption(Chart);
      // 注意这里一定要返回 Chart 实例，否则会影响事件处理等
      return Chart;
```

```
    });
  },
  setOption: function (Chart, isClear = true) {
    if (isClear) { Chart.clear(); }

    var dataList = [];
    var legend = [];
    var found = false;
    var length = 0;
    for (var i = that.data.tableRows.length - 1; i > -1; i--) {
      length = dataList.length;
      found = false;
      for (var k = 0; k < length; k++) {
        if (dataList[k].name == that.data.tableRows[i].recordType) {
          dataList[k].data.push([that.data.tableRows[i].createTime, that.data.tableRows
[i].value]);
          found = true;
          break;
        }
      }

      if (!found) {
        legend.push(that.data.tableRows[i].recordType);
        dataList.push(
          {
            name: that.data.tableRows[i].recordType,
            type: 'line',
            showAllSymbol: true,
            symbolSize: 1,
            data: [[that.data.tableRows[i].createTime, that.data.tableRows[i].value]]
          }
        );
      }
    }

    Chart.setOption(that.getOption(dataList, legend));  //获取新数据

  },
  getOption: function (dataList, legend) {
    var option = {
      color: ["#FF9900", "#99CC33", "#CCCC33", "#FFFF00", "#CC6699", "#3366CC",
"#9933FF", "#FF6666", "#663300", "#993399", "#999966"],
      title: {
        text: '身高体重趋势',
        subtext: '可拖动下面标尺'
      },
```

```
        dataZoom: {
          show: true,
          start: 70
        },
        legend: {
          data: legend,
          width: '65%',
          left: '30%'
        },
        grid: {
          y2: 80
        },
        xAxis: [
          {
            type: 'time',
            splitNumber: 5
          }
        ],
        yAxis: [
          {
            type: 'value'
          }
        ],
        series: dataList
      };
      return option;
    },
    onReady() {
    }
  });
```

上述代码表面看起来比较复杂，实际上只是按照图表的要求，将获取的身高、体重的记录整理成图表要求的格式，赋值到图表的 Option 的指定位置。

至此，本项目的主要内容讲完了。这个项目是为了回顾前面章节涉及的知识点，使读者了解 ASP.NET Core 的基本运行逻辑。为了使项目显得简单、方便理解，其中只引用了几个常用的框架，没有做过多的封装和可能的异常验证。在实际项目中，我们还有很多需要做的工作。在领会了 ASP.NET Core 框架的运行机制之后，再逐步将项目的架构丰富起来，才能更深入地理解为什么要这样做架构。